ELASTIC WAVES IN SOLIDS

WITH APPLICATIONS TO NONDESTRUCTIVE TESTING OF PIPELINES

ELASTIC WAVES IN SOLIDS

WITH APPLICATIONS TO NONDESTRUCTIVE TESTING OF PIPELINES

A.H. Harker, DPhil, CPhys, FInstP

Principal Scientific Officer, Harwell Laboratory

British Gas

Adam Hilger, Bristol and Philadelphia
in association with British Gas plc

First published 1988
under the Adam Hilger imprint by IOP Publishing Ltd
Techno House, Redcliffe Way, Bristol BS1 6NX, England
and 242 Cherry Street, Philadelphia, PA 19106, USA
in association with British Gas plc
Research and Development Division, On Line Inspection Centre, Cramlington, Northumberland

Typeset in Times Roman and Souvenir Gothic at Harwell Laboratory by the author using the TSSD computer typesetting system written by M.J. Hopper.

Printed in Great Britain by
Eyre & Spottiswoode Limited at Grosvenor Press, Portsmouth

ISBN 0 85274 582 6

British Library Cataloguing in Publication Data

Harker, A.H.
Elastic waves in solids with applications to nondestructive testing of pipelines
1. Elastic waves
I. Title II. British Gas plc
531'.33 QA935

ISBN 0–85274–582–6

Foreword

Safe and economical operation of its pipeline network is an important concern to British Gas. For high pressure transmission lines, British Gas has chosen to use nondestructive testing from the inside using vehicle systems which do not affect the normal use of the pipeline. British Gas has already developed a magnetic flux leakage system to identify metal loss defects. This is now well established and is used by the Company to inspect both its own pipelines and those of other pipeline operators around the world. Other inspection systems are being developed, including the elastic wave system to identify longitudinal cracks.

Considerable investment has been made in the development of suitable systems and overcoming the challenges posed by the harsh conditions inside a pipeline. Mechanical engineers have had to design vehicles which can carry complex electronics at a speed of several metres per second, suffering shocks and vibration and negotiating tight bends. Electronics engineers have had to balance their demands for electrical power and data recording against the small space which has been made available to them. Software engineers have had to devise efficient means of interpreting the huge quantities of data which are produced by the detailed inspection of such large surface areas of pipe.

The success of the system, however, depends upon the correct application of the appropriate inspection techniques. In other words, the physics must be right.

For the physics of its system to detect longitudinal cracks, British Gas turned to Harwell. In the mid–1970s Harwell carried out a thorough review of the broad range of nondestructive techniques which might be applied. The final choice was an elastic wave system for which all the basic physics, and some of the engineering, were contracted out to Harwell. As well as drawing on their considerable experience of ultrasonics, new work – both theoretical and experimental – was undertaken. This monograph is an account of the excellent theoretical modelling carried out, placing it in the context of related published work.

Although British Gas has primarily been interested in the application of the results to pipeline inspection, researchers in many other practical uses of elastic waves should find it convenient to have this material gathered together in one volume. I warmly recommend it to them as an erudite, and highly readable, exposition of the subject.

G. Clerehugh, BSc, FEng, FIGasE, MRAeS
British Gas Headquarters Director of Technology

Preface

There are already many books devoted to the theory of elastic waves in solids, so it is natural to question whether another is justified. I believe that there is at present a large gap between the material in standard texts and the area where researchers are currently active, and this book is an attempt to bridge that gap. Most treatises on elastic waves cover the generation of elastic waves by idealized sources, their propagation, reflection and refraction. Some go on to treat the scattering of waves by geometrically simple shapes. Current research, mainly aimed at understanding the scattering of elastic waves in nondestructive testing applications, is concerned with more practical problems: real transducers, complicated scatterers inside awkwardly shaped bodies, noisy anisotropic materials. The tools that have been developed to treat elastic wave scattering in what might be called the real world are often numerical rather than analytical, or rely on a mixture of analysis and computation. The aim of this book is to review some of the modern computational methods, and to show how they relate to exact analytic results, to one another, and to experiment.

In an attempt to keep the book self-contained, it has been necessary to develop some of the standard results of elastic wave theory before moving on to newer developments. This means that it should be suitable for graduate scientists and engineers with a firm background in mechanics and mathematics, but no previous experience of elasticity theory. It should also prove useful for practitioners of nondestructive testing, both for theoreticians as a summary of the present state of the art and for experimentalists as a guide to how much trust to place in theoretical predictions.

I am grateful to British Gas for supporting the writing of this book, and to all my colleagues at Harwell who taught me so much about elastic wave nondestructive testing during the project carried out for British Gas. Particular thanks go to Dr D. Dunford, British Gas On Line Inspection Centre, and to Dr J.A.G. Temple and Ms J.A. Ogilvy, Harwell, for critical readings of a draft of this monograph, to Miss H. Johnson for editorial assistance, and to Mr M.J. Hopper, Harwell, for advice on the use of his typesetting system (TSSD). Despite their good offices, errors and omissions will still remain: for those I alone am responsible.

A.H. Harker
August 1987

Contents

List of figures

List of figures

CHAPTER 1
Introduction

There can be very few industrial installations in which pipework does not play a critical role. For some industries a pipe network is crucial. British Gas plc, for example, relies on about 200 000 kilometres of line to maintain its service to customers. The integrity of the pipes is vital to the industry. These pipes range in diameter from a few centimetres for house-to-house distribution to more than a metre for the main lines of the natural gas transmission system. In all industries there is a penalty involved in pipe failure, and the magnitude of the penalty depends on many factors. In the nuclear industry, pipe failure may lead to a release of radioactivity, and even a slow leak would be unacceptable. In the water industry (for example, Kirby 1984) it is well known that a substantial fraction of the water that leaves the reservoirs is lost through imperfections in the pipe network before it reaches the customer, but it is only when a major leak occurs and disrupts other services that remedial action is necessary. In process plant, pipes may be operating under high pressure, and pipe defects may lead to sudden catastrophic failure of the system: if this leads to the release of poisonous or inflammable material the consequences may be very severe.

The statistics are impressive. A survey of pipelines (Tiratsoo 1983) estimated that there were more than 3 000 000 km of pipelines installed in the world (outside China and the USSR) at the end of 1981, with 21 000 km of energy pipelines under construction and a further 160 000 km projected. Even within the UK there is a very large installed length. The applications include the National Transmission System, owned and operated by British Gas plc, for natural gas, which handled 323 000 000 m^3/day during the winter of 1986/1987.

Against this background, there are obvious economic and social pressures to maintain the integrity of pipeline systems. In order to achieve this, some form of nondestructive testing is necessary. Dick (1977) surveyed the methods available a few years ago, and assessed their capabilities for various categories of defects. At that time most of the nondestructive testing was carried out using hand-held probes, which were moved about the surface of the object to be tested until the operator of the system noted a response on his display – usually a cathode ray tube. Such methods are unsuitable for many applications, for several reasons. Access to the test object may be restricted, as with buried pipework or lines which pass through radioactive areas. The sheer scale of the problem may make manual inspection impracticable: for example the tracks of a railway network. Furthermore, the manual inspection technique may not give enough information. The aim of the inspection is to detect defects which may lead to failure of the system, and a method which is sensitive to metallurgical features which occur in a normal manufacturing process would be of little use. Recent years have therefore seen a shift of emphasis from mere *detection* towards the *identification* of what is seen by nondestructive techniques.

British Gas compared the use of hydrostatic (overload) testing and periodic nondestructive evaluation and concluded that, on both economic and technical grounds, nondestructive evaluation was the better method of assessing the structural integrity of its gas transmission pipelines. This conclusion was supported by studies of defect behaviour and operational experience of the defects which actually occur in gas pipelines. One requirement of the evaluation was that it could be carried out without having to decommission the transmission line – an *on line* inspection system. As a result, two development programmes were begun early in 1974. The first was aimed at detecting areas of metal loss from pipelines, which might result from corrosion or mechanical damage, and used magnetic flux leakage. The second programme was to devise a method for detecting crack-like defects, which might arise from fatigue or stress corrosion processes. The magnetic inspection work led to the introduction of an inspection vehicle for 600 mm pipelines in 1977. Since then vehicles have been developed for an increasing range of pipeline diameters, currently ranging from 200 to 1050 mm. Inspection operations have been carried out both onshore and offshore and in pipelines carrying gas, oil and other products. An inspection service has been provided to pipeline operators around the world, including the British Pipeline Agency in the United Kingdom and the NOVA Corporation in Canada.

The crack detection programme was initiated by British Gas, who placed a contract with Harwell Laboratory and, after a careful review of techniques, elastic wave testing was selected as the most promising. After several years of research and development an inspection vehicle was designed and built: all the basic physics and some of the engineering associated with this system were developed at Harwell. Considerable experience was obtained of operating the prototype vehicle and analysing the data it collected. Elastic wave vehicles have now been deployed in several transmission lines both within the United Kingdom and abroad.

Methods of inspection which use elastic waves have been in use throughout the technological age. A once familiar figure on our railways was the wheel tapper, who used a hammer to set up elastic vibrations of audible frequency in the wheels of rolling stock. In recent years the trend has been towards higher frequencies, and ultrasonic methods are now familiar in many fields. Elastic waves will propagate in solids, liquids and gases, and so the whole range of technological materials is accessible. The frequencies which are available cover an enormous range, but in this book we are concerned with the ultrasonic regime, from about 10^5 to 10^8 Hz. As sound waves in solids travel at speeds of a few mm/μsec, this means that the range of available wavelengths is from centimetres to microns. Graff (1981) traces the modern era of ultrasound back to Langevin in 1915, who was investigating the detection of submarines. Sokolov (1929) and Mühlhäuser (1931) pioneered the use of ultrasound in nondestructive testing. An excellent survey of modern applications of ultrasound in nondestructive evaluation is given by Thompson and Thompson (1985). A firm understanding of the way in which elastic waves propagate in solids and interact with internal features of materials is essential to the selection of appropriate testing procedures. Even the choice of a suitable wavelength from the huge range available is not a trivial task.

The approach adopted in this book is unashamedly theoretical. The aim is to enable the reader to understand ultrasonic testing, and thus to allow him to undertake the design of testing procedures for his own special needs. Of course the level of theory needed in any case will be variable – but only a firm grasp of the principles allows the correct approach to be chosen. A back-of-the-envelope calculation may be adequate when, for example, the wavelength is short and basic ideas of the reflection of rays can be used. More elaborate work is necessary to understand the scattering of a wave from an object whose dimensions are comparable with the wavelength. These levels of theory would correspond to that required by a technician following the introduction to nondestructive testing by Drury (1978) and to that of an expert developing novel testing methods, as reviewed in Szilard (1982) or the series edited by Sharpe (1970–). An intermediate level of understanding would be required for the compendium by Krautkrämer and Krautkrämer (1983).

However elaborate the theoretical techniques become, though, the ultimate aim is the same – to describe what goes on between the insertion of an elastic wave into a solid and its detection and display. The word *between* is important. This book does not cover the theories of piezoelectricity or of electromagnetism which are relevant to the most common kinds of transducer used to generate and detect ultrasonic waves (see Silk 1984). Nor does it cover materials with properties which cannot be described by simple linear elasticity (Hooke's law). This excludes visco-elastic and plastic waves from consideration. Composite materials are not treated explicitly, but this will only exclude a tiny fraction of pipelines. The level of treatment is similar to that in the books on elastic waves by Achenbach (1973), Auld (1973), Graff (1975) and Hudson (1980): reference may be made to these and other standard works to expand on specific points.

CHAPTER 2
Basic ultrasonics

The ultrasonic waves treated in this book arise from the linear equations of elasticity. These are the generalizations to three dimensions of Hooke's law for a linear spring, and assume a simple linear relationship between the distortions of an elastic body and the forces applied to it. For the linear spring, the distortion is expressed in terms of the ratio of the extension to the original length, and the force is a simple tensile or compressive force acting along the length of the spring. For a three-dimensional solid, more complicated measures of distortion (strain) and force (stress) are required.

2.1 Fundamentals of elasticity theory

Strain is defined in terms of the displacements of points inside the body under deformation. Let us suppose that two points at positions $\mathbf{r}(\equiv (r_1, r_2, r_3))$ and $\mathbf{r}+d\mathbf{r}$ are moved by the deformation to $\mathbf{r}'$ and $\mathbf{r}'+d\mathbf{r}'$. Further, suppose that the vector from $\mathbf{r}$ to $\mathbf{r}'$ is $\mathbf{u}$, and that from $\mathbf{r}+d\mathbf{r}$ to $\mathbf{r}'+d\mathbf{r}'$ is $\mathbf{u}+d\mathbf{u}$. Figure 1 shows the relationships between these vectors.

Figure 1

Displacements within a solid body under deformation, showing the variation of displacement **u** with position **r**.

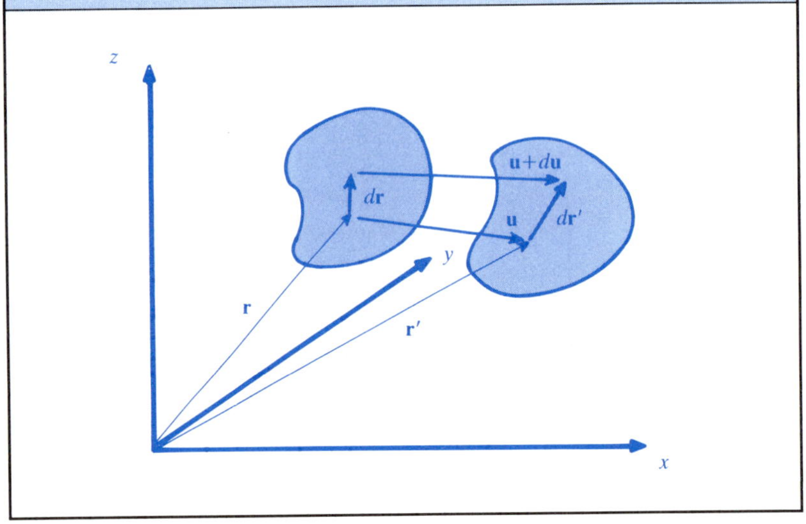

Now we have, in component notation,

$$u_i = r'_i - r_i, \tag{2.1}$$

but

$$dr_i = \sum_j \frac{\partial r'_i}{\partial r_j} dr_j \tag{2.2}$$

so that

$$du_i = \sum_j \frac{\partial u_i}{\partial r_j} dr_j, \tag{2.3}$$

which we write in the abbreviated notation

$$du_i = u_{i,j} dr_j, \tag{2.4}$$

that is, we adopt a comma notation within the subscript to denote differentiation and we use the convention that a repeated index is to be summed over. If we separate this expression into symmetric and antisymmetric components,

$$du_i = \tfrac{1}{2}(u_{i,j}+u_{j,i})dx_j + \tfrac{1}{2}(u_{i,j}-u_{j,i})dx_j, \tag{2.5}$$

the components correspond to local distortion effects, described by the strain e_{ij} and local rigid-body rotations ω_{ij}:

$$e_{ij} \equiv \tfrac{1}{2}(u_{i,j}+u_{j,i}) \tag{2.6}$$

$$\omega_{ij} \equiv \tfrac{1}{2}(u_{i,j}-u_{j,i}). \tag{2.7}$$

The strain tensor e_{ij} is symmetric: this means that it may be diagonalized by rotating the coordinate system into the principal axis system. In the principal axis system the diagonal elements are known as the principal strains. As long as the gradients of **u** are small, the order in which the strain and the rotation are applied is immaterial – the operators commute in the limit of infinitesimal distortions.

We now have to describe the forces acting on and within the body. The picture we have is as follows. If, within the body, we were to form a small crack, then by applying appropriate tractions to the crack surfaces we could mimic the effect of the material which used to be in contact at those surfaces. On an arbitrary surface element $d\mathbf{S} \equiv (dS_1, dS_2, dS_3)$ within the body there will be a traction $d\mathbf{t} \equiv (dt_1, dt_2, dt_3)$. The direction of the tractive force may make any angle with the vector defining the surface, and the two are related by the stress tensor τ_{ij}:

$$dt_i = \tau_{ij} dS_j. \tag{2.8}$$

That $\boldsymbol{\tau}$ is indeed a tensor is proved by, for example, Nye (1957, page 87) and Hudson (1980, page 9).

The momentum of the body per unit volume is the product of the density ρ and the local velocity. If the external forces are represented by tractions $\mathbf{t} \equiv (t_1, t_2, t_3)$ on the surface S and body forces acting on unit mass $\mathbf{f} \equiv (f_1, f_2, f_3)$, conservation of momentum requires that

$$\frac{D}{Dt}\int_V \rho u_{i,t} dV = \int_S dt_i + \int_V \rho f_i dV \tag{2.9}$$

where we again use the comma notation to denote differentiation, now with respect to time. Using Gauss' theorem and rearranging, this gives

$$\int_V (\rho u_{i,tt} - \tau_{ji,j} - \rho f_i) dV = 0, \tag{2.10}$$

or, as V is arbitrary,

$$\rho u_{i,tt} - \tau_{ji,j} - \rho f_i = 0. \tag{2.11}$$

Conservation of energy, in the context of an elastic body, requires that the rate of change of kinetic energy plus potential energy is equal to the rate at which work is done by external forces. The kinetic energy per unit volume is the product of the density ρ and the square of the local velocity. If $U(\mathbf{r})$ is the internal energy of the body per unit volume and E is the total internal energy then

$$E = \int_V U \, dV \tag{2.12}$$

and the kinetic energy per unit volume is

$$K = \tfrac{1}{2}\rho u_{i,t} u_{i,t}. \tag{2.13}$$

Then conservation of energy requires that

$$\frac{D}{Dt}\int_V (\tfrac{1}{2}\rho u_{i,t} u_{i,t} + U) dV = \int_S t_i u_{i,t} \, dS + \int_V \rho f_i u_{i,t} \, dV \tag{2.14}$$

which may be converted, using the divergence theorem in the surface integral, to

$$\int_V (\rho u_{i,t} u_{i,tt} + U_{,t}) dV = \int_V (\tau_{ij} u_{i,jt} + f_i u_{i,t} \rho) dV \tag{2.15}$$

or

$$\int_V \{(\rho u_{i,tt} - \tau_{ji,t} - \rho f_i) u_{i,t} + (U_{,t} - \tau_{ji} u_{i,jt})\} dV = 0. \tag{2.16}$$

Now the first term in this integral is zero from the conservation of momentum. Furthermore, $u_{i,j}$ is the sum of the symmetric tensor e_{ij} and the antisymmetric ω_{ij}. The product of the antisymmetric tensor with the symmetric τ_{ij} is zero, so we have

$$U_{,t} = \tau_{ji} e_{ji,t}. \tag{2.17}$$

If the material is perfectly elastic, in the sense that there are no mechanisms for dissipating energy, we may assume (Green 1839) that the strain energy of the material is a function of the strain only (and not, for example, dependent on the strain rate). We can derive a relationship between the stress and the strain as follows. We assume that

$$U = U(e_{ij}) \tag{2.18}$$

so that

$$U_{,t} = \frac{\partial U}{\partial e_{ij}} e_{ij,t}. \tag{2.19}$$

Using the energy conservation relation, this gives

$$\tau_{ij} = \frac{\partial U}{\partial e_{ij}}. \tag{2.20}$$

Alternatively (Cauchy 1828), we may take the stress to be determined only by the instantaneous value of the strain:

$$\tau_{ij} = \tau_{ij}(e_{ij}). \tag{2.21}$$

In each case we may assume that the strains are small enough for an expansion in powers of strain to be appropriate, and then, taking the unstrained state as the reference state, the stress is linearly related to the strain:

$$\tau_{ij} = c_{ijkl} e_{kl}. \tag{2.22}$$

Equation (2.22) is known as the stress–strain relationship, or the constitutive equation. The strain energy density is then given by

$$U = \tfrac{1}{2} c_{ijkl} e_{ij} e_{kl}. \tag{2.23}$$

The elastic constant matrix c_{ijkl} may be shown (Nye 1957, page 134; Hudson 1980, page 15) to be a tensor. It has several symmetries:

$$c_{ijkl} = c_{klij} = c_{jikl} = c_{ijlk}, \tag{2.24}$$

so that only 21 of its 81 elements are independent. In fact Cauchy's theory for elastic bodies was rather more restrictive than this, being based on a microscopic model of a solid made up of microscopic masses with pairwise interactions. This leads to only 15 independent elements, but experiments have shown the theory with 21 elements to be correct. Further discussion of this and related points is given by Huntington (1958). The conclusion for practical materials is, however, that Green's assumption seems to be valid: such materials are called *hyperelastic*. If the strain energy is not fully defined by the strain alone the symmetry

$$c_{ijkl} = c_{klij} \tag{2.25}$$

need not hold, and the elastic constant tensor has 36 independent elements. Such materials are *elastic* but not *hyperelastic*.

A modified notation is often found for the elastic constants, in which the symmetry of the stress and strain tensors is explicitly reflected. In this scheme one writes

$$\tau_1 = \tau_{11} \quad \tau_2 = \tau_{22} \quad \tau_3 = \tau_{33} \quad \tau_4 = \tau_{23} \quad \tau_5 = \tau_{31} \quad \tau_6 = \tau_{12} \tag{2.26}$$

and similarly for the strains. Then c_{ijkl} is written as the 6 by 6 matrix c_{mn}. We shall use the full four-index notation throughout this book.

Further symmetries of the material may further reduce the number of independent elastic constants. A full treatment is given by Sokolnikov (1956), and the theory of elastic waves in crystals is treated by Fedorov (1968). In many applications, however, there are no preferred directions within the material and the elastic constant tensor is isotropic. We shall show below that this is true of polycrystalline materials if the grain size is small enough (see Chapter 4). In that case there are only two independent parameters, λ and μ, called the Lamé constants (Lamé 1852), and we may write

$$c_{ijkl} = \lambda\delta_{ij}\delta_{kl} + \mu\delta_{ik}\delta_{jl} + \mu\delta_{il}\delta_{jk}. \tag{2.27}$$

The stress–strain relationship then reduces to

$$\tau_{ij} = \lambda e_{kk}\delta_{ij} + 2\mu e_{ij} \tag{2.28}$$

and the strain energy density to

$$U = \tfrac{1}{2}\lambda e_{ii}e_{jj} + \mu e_{ij}e_{ij}. \tag{2.29}$$

The elastic constants of isotropic materials are often quoted in terms of the bulk modulus B, Young's modulus Y and Poisson's ratio ν. Poisson's ratio is the ratio between the transverse contraction and the longitudinal extension of a rod subjected to a pure stretching force. Some authors use different notations, denoting Young's modulus by E and the bulk modulus by K. The relationships between the constants are as follows:

$$B = \lambda + \tfrac{2}{3}\mu \tag{2.30}$$

$$Y = \frac{\mu(3\lambda + 2\mu)}{\lambda + \mu} \tag{2.31}$$

$$\nu = \frac{\lambda}{2(\lambda+\mu)}. \quad (2.32)$$

Inverting these formulae,

$$\lambda = \frac{Y\nu}{(1+\nu)(1-2\nu)} \quad (2.33)$$

$$\mu = \frac{Y}{2(1+\nu)} \quad (2.34)$$

$$B = \frac{Y}{3(1-2\nu)}, \quad (2.35)$$

which shows that $\nu<\frac{1}{2}$ or the material would contract under pressure, and that $\nu>-1$ or the energy density would not be positive definite. For a fluid, $\nu = \frac{1}{2}$. Most structural materials are polycrystalline, and approximate to isotropic materials. For such materials negative values of ν have not been observed. Very low values of ν can occur: beryllium metal has a value of 0.05. For single crystals effective values of ν can be defined, which are direction-dependent, and single crystals of silica under certain stress conditions (Kittinger, Tichý and Bertagnolli 1981) have a negative ν. It is easy to see that a negative ν is possible, especially for a material such as silica which is made up of quite rigid tetrahedral units. In extending a rod it may be necessary to rotate the rigid units, so increasing their cross-section perpendicular to the direction of stretching. Love (1944) cites a pyrites crystal with a negative Poisson's ratio, and suggests that the crystal may have been twinned. Other cases of direction-dependent Poisson's ratio are described by Gunton and Saunders (1975): they find some negative values. Lakes (1987) describes polymer foams with reentrant structures which have Poisson's ratios as low as –0.7. More commonly, however, ν lies in the range $\frac{1}{4}$ to $\frac{1}{3}$.

It should be noted that the expressions we have derived so far hold only in the limit of small strains. In more general cases the elastic energy is represented as an expansion to higher order in the strains:

$$U = \tfrac{1}{2}c_{ijkl}e_{ij}e_{kl} + \tfrac{1}{6}c_{ijklmn}e_{ij}e_{kl}e_{mn} + \ldots, \quad (2.36)$$

and c_{ijkl} is more properly known as the second order elastic constant tensor. The higher order elastic constant tensors continue with c_{ijklmn}, the third order tensor. The higher order tensors are generally of no concern in ultrasonic nondestructive testing because the amplitudes of the waves used, and hence the strains, are extremely small. There is one exception, however. It is sometimes desirable to determine the residual stress in a manufactured specimen, that is, the stress that has been built in by such mechanisms as differential cooling, rolling and pressing. In such situations the static strain field may affect the elastic wave velocities in a measurable way: a review of the experimental situation is given by Allen *et al.* (1982), and Pao, Sachse and Fukuoka (1984) have recently summarized the theoretical position. One particular problem (Sayers 1982; Allen and Sayers 1984) is that the processes which introduce residual stress also often build directional character into the material, in the form of a texture in the grain structure. Suitable combinations of measurements can eliminate the effects of texture: Pritchard (1987) assesses the experimental situation. Residual stresses can have dramatic effects, even though the changes in wave velocity are typically less then 0.1 per cent: for an example see Smith, Richardson and van der Merwe (1987).

2.2 The wave equation: differential form

When the conservation of momentum condition is combined with the constitutive equation (stress–strain relationship) the equation of motion of the solid becomes

$$(c_{ijkl}u_{k,l})_{,j}+\rho f_i = \rho u_{i,tt} \tag{2.37}$$

which reduces, for an isotropic solid, to

$$(\lambda u_{k,k})_{,i}+(\mu(u_{i,j}+u_{j,i}))_{,j}+\rho f_i = \rho u_{i,tt}. \tag{2.38}$$

If, further, the material is homogeneous, the elastic constants λ and μ will be independent of position. Then

$$(\lambda+\mu)u_{j,ji}+\mu u_{i,jj}+\rho f_i = \rho u_{i,tt}, \tag{2.39}$$

or, in vector form,

$$(\lambda+\mu)\boldsymbol{\nabla}\boldsymbol{\nabla}\cdot\mathbf{u}+\mu\nabla^2\mathbf{u}+\rho\mathbf{f} = \rho\frac{\partial^2\mathbf{u}}{\partial t^2}. \tag{2.40}$$

This is the equation of dynamic elasticity with which we shall be concerned throughout this book. We shall not dwell too long on the mathematical fundamentals of elastodynamics: the finer points are covered by Wheeler and Sternberg (1968) and Knops and Payne (1971). The important point is that the equation is linear, and this allows us (Sokolnikov 1956) to investigate the uniqueness of the solution. Suppose that there exist two independent solutions to a given elastodynamic problem, denoted by $(\mathbf{u}',\boldsymbol{\tau}',\mathbf{e}')$ and $(\mathbf{u}'',\boldsymbol{\tau}'',\mathbf{e}'')$. Then because the elastodynamic equation is linear, the difference of these two solutions must also be a solution: let this difference be $(\mathbf{u},\boldsymbol{\tau},\mathbf{e})$. It is clear that this difference must satisfy the homogeneous equation with $\mathbf{f} = 0$. Now, from the conservation of energy, we know that

$$\frac{D}{Dt}\int_V(\tfrac{1}{2}\rho u_{i,t}u_{i,t}+U)dV = \int_S t_i u_{i,t}\, dS \tag{2.41}$$

which may be integrated with respect to time to give

$$\int_V [\tfrac{1}{2}\rho u_{i,t} u_{i,t} + U]_0^t dV = \int_0^t \int_S t_i u_{i,t} \, dS \, dt. \tag{2.42}$$

Now the integral on the left-hand side contains the differences between the values at times 0 and t of the kinetic and potential energies: these are both positive definite quantities. The initial values of both of them are zero, based as they are on the difference between the two solutions, provided that the initial conditions specify $\mathbf{u}$ and $\mathbf{u}_{,t}$ throughout the volume V. If the solution is to be unique, this means that the final sum of the kinetic and potential energies based on the differences must also be zero. This will be so if the term on the right-hand side is also zero. This shows that the solution will be unique if boundary conditions are specified on the surface S which specify $\mathbf{u}$, $\mathbf{t}$, or a suitable combination.

We have thus seen that to solve for the motion of the solid, giving the displacement $\mathbf{u}$ at every point in the body as a function of time, we must specify suitable boundary conditions. In practical cases the boundary conditions are supplied by applied stresses – cases with totally rigid boundaries (and therefore zero displacements) are good approximations in some acoustic problems, but not for most elastodynamic cases. If we rewrite the integrand $t_i u_{i,t}$ in terms of stresses and displacements, using a local coordinate system based on the normal to the surface n and two orthogonal vectors in the surface a,b, we can see that we must specify τ_{nn} or u_n and τ_{na} or u_a and τ_{nb} or u_b over the whole surface S, as well as the initial values of $\mathbf{u}$ and $\mathbf{u}_{,t}$ throughout the volume V.

Many practical problems may be treated as though they were two-dimensional. For example, a crack partly penetrating a thick stretched plate will give rise to a stress field which varies much more rapidly in a plane perpendicular to the crack than it does along the length of the crack. Two-dimensional problems are characterized as plane strain, anti-plane strain, plane stress or anti-plane stress. The example we have just given is one of plane (or anti-plane) strain.

Figure 2 shows the limiting case of an infinite crack: the geometry of the system is constant along the x_2 direction. Plane strain is defined as having displacement $u_2 = 0$. This does not mean that the stress τ_{22} is zero, but it does imply that τ_{12} and τ_{23} are zero and that τ_{22} is related to the other stresses by

$$\tau_{22} = -\nu(\tau_{11} + \tau_{33}). \tag{2.43}$$

In anti-plane strain, on the other hand, only the displacements parallel to the x_2 axis are non-zero:

$$u_1 = 0 \tag{2.44}$$

$$u_3 = 0 \tag{2.45}$$

Figure 2

Plane and anti-plane strain of a deformed body which exhibits no variation in the x_2 direction: under plane strain the displacement u_2 is zero, and under anti-plane strain the displacements u_1 and u_3 are zero.

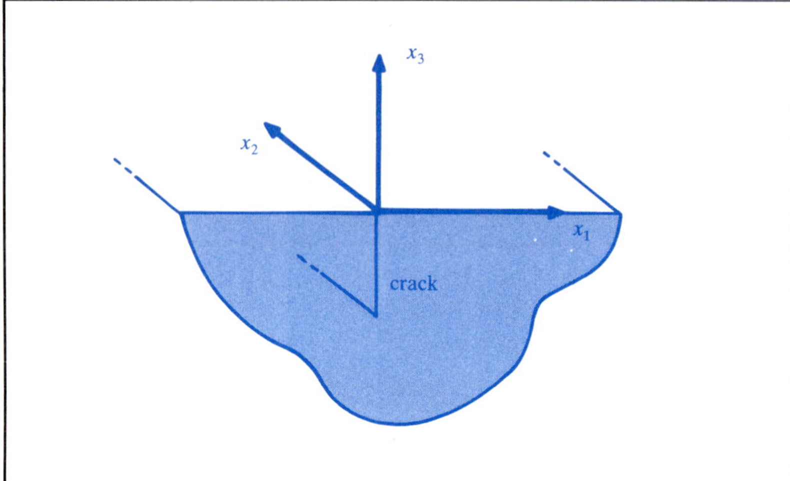

and the stresses

$$\tau_{11} = \tau_{22} = \tau_{33} = \tau_{13} = 0. \tag{2.46}$$

Plane stress conditions will not concern us here. They are applicable to thin elastic plates under the influence of forces applied in their own plane. The equations involved apply to the averages of stresses and displacements across the thickness of the plate, and are derived (Poisson 1829; Cauchy 1828) by expanding the displacements in powers of the plate thickness and truncating the series.

2.2.1 Potentials, compression and shear waves

The wave equation expressed in terms of displacements, equation (2.39),

$$(\lambda+\mu)u_{j,ji}+\mu u_{i,jj}+\rho f_i = \rho u_{i,tt},$$

is rather unwieldy, as it couples the various components of displacement. A much more convenient method is to introduce potentials, and to relate the displacements to the derivatives of the potentials. In general one may write (for a discussion see Sternberg 1960)

$$\mathbf{u} = \boldsymbol{\nabla}\phi+\boldsymbol{\nabla}_{\wedge}\boldsymbol{\psi}. \tag{2.47}$$

As this relates the three components of $\mathbf{u}$ to four functions (ϕ and the three components of $\boldsymbol{\psi}$), an additional condition is required. The conventional choice is

$$\nabla \cdot \boldsymbol{\psi} = 0. \tag{2.48}$$

This is known as the Helmholtz resolution of the vector **u**. To complete the separation of the equations, we resolve the inhomogeneous term (the driving force **f**) using

$$\mathbf{f} = \nabla \alpha + \nabla_{\wedge} \boldsymbol{\beta} \tag{2.49}$$

and

$$\nabla \cdot \boldsymbol{\beta} = 0. \tag{2.50}$$

Substituting into the wave equation and grouping terms,

$$\nabla\left\{(\lambda+2\mu)\nabla^2\phi+\rho\alpha-\rho\phi_{,tt}\right\}+\nabla_{\wedge}\left\{\mu\nabla^2\boldsymbol{\psi}+\rho\boldsymbol{\beta}-\rho\boldsymbol{\psi}_{,tt}\right\} = 0, \tag{2.51}$$

and this equation will be satisfied if each of the terms in braces is zero:

$$(\lambda+2\mu)\nabla^2\phi+\rho\alpha = \rho\phi_{,tt} \tag{2.52}$$

$$\mu\nabla^2\boldsymbol{\psi}+\rho\boldsymbol{\beta} = \rho\boldsymbol{\psi}_{,tt}. \tag{2.53}$$

Now each of these equations has the characteristic form of a wave equation, and the potentials ϕ and $\boldsymbol{\psi}$ propagate with characteristic velocities

$$c_p = \sqrt{\frac{\lambda+2\mu}{\rho}} \tag{2.54}$$

and

$$c_s = \sqrt{\frac{\mu}{\rho}}. \tag{2.55}$$

For typical solids, c_p is approximately twice c_s. The subscripts p and s stem from seismology, where the two disturbances detected at a site some way from an earthquake were named primary and secondary. For reasons we shall discuss below, the subscripts may also be interpreted as standing for *push* and *shake,* and the wave motions are also known as *longitudinal* and *transverse* or *compression* and *shear.* The ratio of the two wave speeds is entirely determined by the Poisson's ratio of the solid, so that

$$\frac{c_p}{c_s} = \sqrt{\frac{2-2\nu}{1-2\nu}} \tag{2.56}$$

or, inversely,

$$\nu = \frac{(c_p/c_s)^2-2}{2((c_p/c_s)^2-1)}. \tag{2.57}$$

The separation into two wave modes considerably simplifies the analysis. The analytical properties of the simple wave equations are well understood, and solutions in terms of standard functions are known. The compression and shear waves are, however, coupled at boundaries: this introduces the phenomenon of mode conversion which we shall describe later.

2.2.2 Time-harmonic solutions

The solution of a scalar wave equation

$$c_p^2 \nabla^2 \phi = \phi_{,tt} \tag{2.58}$$

may be written

$$\phi = Af(\hat{\mathbf{k}} \cdot \mathbf{r} - c_p t), \tag{2.59}$$

where A is the amplitude of the wave, f is an arbitrary function, and $\hat{\mathbf{k}}$ is known as the wave normal. A rather more familiar form is

$$\phi = Ag(\mathbf{k}_p \cdot \mathbf{r} - \omega t) \tag{2.60}$$

in which we have used the wavevector $\mathbf{k}_p$ and angular frequency ω. The quantity $(\mathbf{k}_p \cdot \mathbf{r} - \omega t)$ is known as the phase of the wave, from which it is evident that the wave normal (and thus the wavevector) is the normal to surfaces of constant phase. The corresponding displacement is

$$\mathbf{u} = A\mathbf{k}_p g'(\mathbf{k}_p \cdot \mathbf{r} - \omega t), \tag{2.61}$$

showing that for this compressional wave the displacement of the material is parallel to the wavevector. This form of motion corresponds to the familiar acoustic wave, where there are alternating regions of compression and rarefaction.

To extend the analysis to the vector potential $\boldsymbol{\psi}$ it is only necessary to replace the constant A with a vector: this may be written $\mathbf{B}$, where $|B|$ is the amplitude and $\mathbf{B}/|B|$ is a unit vector specifying the polarization. Then we have

$$\boldsymbol{\psi} = \mathbf{B}h(\mathbf{k}_s \cdot \mathbf{r} - \omega t) \tag{2.62}$$

giving

$$\mathbf{u} = \mathbf{k}_{s\wedge} \mathbf{B}h'(\mathbf{k}_s \cdot \mathbf{r} - \omega t). \tag{2.63}$$

This shows that the displacement in this wave is perpendicular to the wavevector, justifying the name transverse or shear.

For anisotropic media Stroh (1962) writes the solution of equation (2.39) in the form

$$\mathbf{u} = \mathbf{A}f(\mathbf{m} \cdot \mathbf{x} + p\mathbf{n} \cdot \mathbf{x} - ct), \tag{2.64}$$

where $\mathbf{m}$ and $\mathbf{n}$ are orthornormal vectors, c is a constant speed, and p

and **A** are determined by solving a six-dimensional eigenvalue problem for a matrix defined in terms of **m**, **n** and ρc^2. The bulk waves correspond to solutions in which p is real, and travel with speed c. The variations of propagation speed with direction are conveniently represented in terms of slowness surfaces, in which the inverse of the wave speed is plotted in polar form. A clear introduction to the use of slowness diagrams is given by Musgrave (1978).

The speeds c_p and c_s are known as the phase velocities. From the form of the equations it is clear that they are the rates at which particular points on the wave train, for example the peaks of a harmonic wave, pass a given point. There are two other speeds which it is important to consider: these are the group velocity v_g, at which a modulation or wave packet travels, and the energy transport velocity v_e. The group velocity is most simply derived (following Stokes 1876) by considering a one-dimensional wave constructed by superposing two waves with slightly different values of ω and k:

$$\begin{aligned} g(x,t) &= \cos(kx-\omega t) + \cos[(k+\delta k)x - (\omega+\delta\omega)t] \\ &= 2\cos[(k+\tfrac{1}{2}\delta k)x-(\omega+\tfrac{1}{2}\delta\omega)t]\,\cos\left[\frac{\delta k}{2}x-\frac{\delta\omega}{2}t\right]. \end{aligned} \tag{2.65}$$

The resulting waveform is the product of the carrier with a propagation speed

$$v_p = \frac{\omega+\frac{1}{2}\delta\omega}{k+\frac{1}{2}\delta k} \rightarrow \frac{\omega}{k}, \tag{2.66}$$

known as the phase velocity, and the envelope with a speed

$$v_g = \frac{\delta\omega}{\delta k} \rightarrow \frac{\partial\omega}{\partial k}. \tag{2.67}$$

The three-dimensional analogue has

$$(v_{gx}, v_{gy}, v_{gz}) = \left(\frac{\partial\omega}{\partial k_x}, \frac{\partial\omega}{\partial k_y}, \frac{\partial\omega}{\partial k_z}\right). \tag{2.68}$$

It should be emphasized that the group velocity v_g, like the phase velocities, is in general dependent on the direction of propagation. Only in the case of isotropic elasticity is it independent of direction.

The energy transport velocity is defined as the ratio between the time-averaged power flow density and the time-averaged stored energy density in the elastic wave. The energy flow through an element of area $d\mathbf{S}$ may be found from the rate at which work is done by one side of the area on the other, the product of the strain rate and the traction, as shown before:

$$dF = \mathbf{u}_{,t}\cdot\boldsymbol{\tau}\cdot d\mathbf{S} \tag{2.69}$$

so that the energy flux per unit area is

$$\mathbf{F} = \mathbf{u}_{,t} \cdot \boldsymbol{\tau}. \tag{2.70}$$

The modulus of **F** is F, known as the intensity. A little care is necessary when using a complex representation of the wavefield, in which a sinusoidal wave is represented as $e^{i(kx-\omega t)}$. In calculating the energy, it is the *real* parts of the stress and the strain which should be multiplied together, so that if we use an asterisk to denote complex conjugation we have

$$\mathbf{F} = \tfrac{1}{4}(\mathbf{u}_{,t} + (\mathbf{u}_{,t})^*) \cdot (\boldsymbol{\tau} + \boldsymbol{\tau}^*). \tag{2.71}$$

The energy density is the sum of the kinetic and strain energies, which are given, as in equations (2.13) and (2.23), by

$$K = \tfrac{1}{2}\rho u_{i,t} u_{i,t} \tag{2.72}$$

and

$$U = \tfrac{1}{2} c_{ijkl} e_{ij} e_{kl}. \tag{2.73}$$

In most cases, the rate of energy transport is equal to the group velocity, and several explicit verifications were given by Biot (1957). The main exception is when the medium is dissipative, and energy is absorbed. Even then, if the dissipation is small the concept of group velocity is still useful, and it is possible to describe the motion of a wave packet in terms of its group velocity and a decay of amplitude with time. Brillouin (1914) and Baerwald (1930) have also introduced the signal velocity: in essence this is defined in terms of the time at which the amplitude of a pulse reaches a certain fraction of its peak value. This differs from the group velocity only in highly dispersive systems, for example, near a resonant absorption. Asymptotically, for narrow resonances, the signal velocity is equal to the energy transport velocity (Brillouin 1960, page 119). For very dispersive systems, however, there are problems in defining useful values for these velocities: a discussion is given by Elices and Garcia–Moliner (1968). General results on energy transport velocity are most conveniently obtained by using a Lagrangian or Hamiltonian formulation of the wave motion: for example Lighthill (1965), Tolstoy (1973, Chapter 1) and Achenbach (1973, Section 6.5).

2.2.3 Impedance and the Kramers–Kronig relations

The impedance concept is sometimes used in elastodynamic work, especially by workers who cut their teeth on electromagnetic problems. The impedance gives the response of the system to an external stimulus. Impedance is particularly useful in reflection and refraction problems, where it can be used to quantify the discontinuity ('mismatch') at the interface.

By analogy with electric circuit theory, where impedance Z is defined as the ratio of driving voltage to current flow, one can define an elastodynamic impedance as the ratio of force to velocity. There are different interpretations of the term 'force', which may be taken to be a stress or a traction, but in general the impedance $\mathbf{Z}$ is a matrix quantity.

Morse and Feshbach (1953) consider a simple harmonic wave with angular frequency ω, with a velocity

$$\mathbf{u}_{,t} = i\omega\mathbf{u} \tag{2.74}$$

for displacement $\mathbf{u}$. The traction on a unit area normal to the unit vector $\mathbf{n}$ is then

$$\begin{aligned} t_i &= n_j c_{ijkl} e_{kl} \\ &= i\omega Z_{ij} u_j, \end{aligned} \tag{2.75}$$

which defines the impedance matrix $\mathbf{Z}$. Other workers (for example, de Hoop 1985) take the impedance to be the ratio of stress to velocity. For a harmonic wave, then,

$$\tau_{ij} = i\omega Z'_{ijk} u_k. \tag{2.76}$$

Here we use the first definition of impedance, equation (2.75). For a compression wave in an isotropic medium, if $\mathbf{n}$ is parallel to $\mathbf{k}_p$, the traction $\mathbf{t}$ is parallel to the velocity $\mathbf{u}_{,t}$, and the impedance matrix is simply

$$Z^p_{ij} = \rho c_p \delta_{ij}, \tag{2.77}$$

that is, ρc_p times a unit matrix. For shear waves, the impedance matrix is again a multiple of a unit matrix:

$$Z^s_{ij} = \rho c_s \delta_{ij}. \tag{2.78}$$

In this case, both the traction and the velocity are perpendicular to the surface. In anisotropic media, the impedance matrices can take more complicated forms.

Elastodynamic systems, in common with all linear physical systems, are subject to the Kramers–Kronig relations (Kronig and Kramers 1928; Landau and Lifshitz 1958). These relations are derived from the fact that a physical system must be causal – that is, a response must occur later in time than the stimulus from which it arises. The Kramers–Kronig expressions relate the real and imaginary parts of a response function. The response function is generally taken to be a generalized susceptibility, which is the constant of proportionality between a force and a displacement. In the elastic wave case we may define a matrix $\mathbf{A}$ by

$$u_i = A_{ij} t_j, \tag{2.79}$$

that is,

$$\mathbf{A} = (i\omega\mathbf{Z})^{-1}. \tag{2.80}$$

If $\mathbf{A}$ is separated into its real and imaginary parts

$$\mathbf{A} = \mathbf{B}+i\mathbf{C}, \tag{2.81}$$

then the Kramers–Kronig relations may be written as

$$\mathbf{B}(\omega) = \mathbf{B}_\infty + \mathrm{P}\int_{-\infty}^{\infty} \frac{d\omega'}{\pi} \frac{\mathbf{C}(\omega')}{\omega'-\omega} \tag{2.82}$$

$$\mathbf{C}(\omega) = -\mathrm{P}\int_{-\infty}^{\infty} \frac{d\omega'}{\pi} \frac{\mathbf{B}(\omega')-\mathbf{B}_\infty}{\omega'-\omega}. \tag{2.83}$$

Here $\mathbf{B}_\infty$ is the limiting value of the real part of $\mathbf{A}$ at infinite frequency.

These equations show that if there is any frequency dependence of the real part of the response then there will be a non-zero imaginary part at some frequency. In particular, this means that for elastic waves dispersion (variation of wave velocity with frequency) must always be accompanied by attenuation (loss of amplitude with distance travelled). The Kramers–Kronig relations can be useful in analysing the physical properties of materials (for example, O'Donnell, Jaynes and Miller 1981).

2.2.4 Scattering cross-sections

In calculations for nondestructive testing applications, what is generally of interest is the cross-section of a particular flaw for scattering. Here one may define a total cross-section, which is the ratio of the average power scattered in all directions to the average incident intensity:

$$\sigma(\omega) \equiv \frac{\langle \mathrm{P}^s \rangle}{\langle F^i \rangle}. \tag{2.84}$$

Here the angled brackets $\langle\,\rangle$ denote averages over time – in this case averages over one period of a single-frequency wave – and the superscripts i and s refer to the incident and scattered fields. The total scattered power P^s is the integral of the scattered flux over a surface surrounding the scatterer:

$$\mathrm{P}^s = \int_S \mathbf{u}^s_{,t} \cdot \boldsymbol{\tau}^s \cdot d\mathbf{S}. \tag{2.85}$$

The differential cross-section $d\sigma/d\Omega$, where $d\Omega$ is a solid angle element, gives the distribution of scattered power into different directions. At a distance r from the scatterer in the direction $\hat{\mathbf{r}}$ an element of area $d\mathbf{S}$ may be written as

$$d\mathbf{S} = r^2 d\Omega\hat{\mathbf{r}} \tag{2.86}$$

so that

$$\frac{d\sigma}{d\Omega} = \frac{\langle r^2 \mathbf{u}^s_{,t} \cdot \boldsymbol{\tau}^s \cdot \hat{\mathbf{r}} \rangle}{\langle F^i \rangle}. \tag{2.87}$$

In principle, equation (2.87) is to be evaluated in the limit $r \to \infty$.

2.3 Boundary conditions, reflection and refraction

In the previous section, we treated solutions of the elastodynamic equations in an infinite, homogeneous medium, and showed that the compression and shear waves could be treated separately. In general, however, the presence of discontinuities destroys this separation. There are two classes of effect: the phenomena of reflection, refraction and mode conversion which we treat in this section and the special classes of guided and surface waves which we describe in the next section.

The theory of the reflection and refraction of ultrasound at oblique incidence on plane surfaces has been treated in some research papers as well as in most standard texts: for examples see Thomson (1950), Brekhovskikh (1960) and Folds and Loggins (1977). At each interface the problem is simply one of calculating the coefficients of reflected and transmitted waves so as to satisfy the appropriate boundary conditions. At a bonded junction between two media, for example, we have to match the displacements and stresses across the interface. At a free surface the displacements are not specified, but all the components of the traction will be zero. For practical purposes, there are three considerations: changes of amplitude, mode conversions, and changes of phase. It is obvious that if some ultrasonic energy leaks out of the specimen at an interface this will affect the amplitude of the signal which is observed. Rather more subtle is the effect of mode conversion, by which all the energy may remain in the specimen but be converted to a mode which will not give a significant signal at the receiving transducer. Furthermore, as we shall show below, important effects of beam displacement and pulse shape changes arise from the phase changes.

If one has a means of calculating the transmission and reflection of ultrasound at an interface, or at a layer of one medium between two others, one can understand many important phenomena. Typical problems which can be addressed concern the efficiency with which ultrasound can be injected into a material, and the optimum angle for a given transducer material; the effect that a material filling a crack might have on the reflectivity of the crack, and hence on the detectability of crack-like flaws; the choice of a coupling material to optimize the transfer of ultrasound between the transducer and the test specimen; and the effect of layers, for example of coupling material, on the frequency content of the ultrasonic signals.

Figure 3

The geometry of an interface, defining the angle θ used in the text.

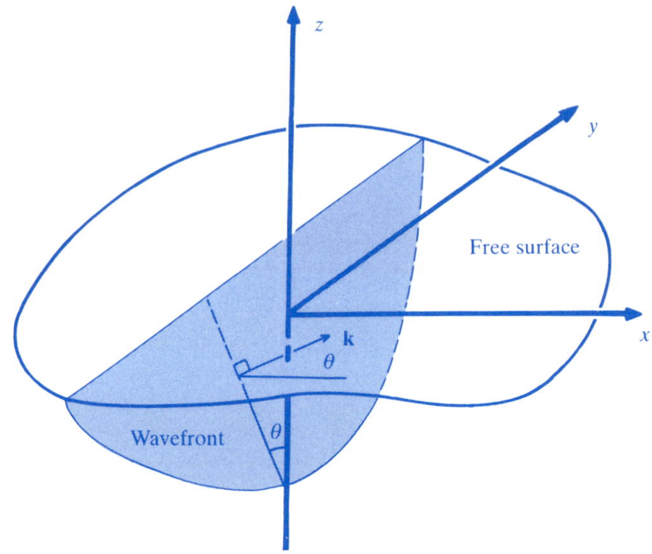

If we consider an interface of infinite extent in the yz plane, and an incident plane wave with its wavevector in the xz plane making an angle θ with the x axis (Figure 3), then we have two possible cases. In the first, and simplest, the plane wave is a shear wave giving rise to displacements in the y direction, that is, parallel to the interface. In this case no mode conversion takes place and a shear wave polarized parallel to the interface is reflected and a similar wave transmitted. This polarization is designated SH: this notation arises from seismology, where the interface is the horizontal surface of the earth, and the displacement parallel to that interface is horizontal. The second case has either a compression (P) wave or a shear wave polarized in the xz plane (SV) incident: again the notation arises from seismology, with a vertical displacement. In general this will result in reflected and transmitted P and SV waves.

For plane waves of frequency ω the potentials for the incident waves may be written as in terms of complex exponentials with amplitude factors for P, SV and SH waves of A^i, B^i and C^i: note that these amplitudes have the dimensions of area. The incident waves are as follows:

P wave

$$\phi^i = A^i e^{i(k_{px}x + k_{pz}z - \omega t)} \tag{2.88}$$

SV wave

$$\psi^{i,\mathrm{SV}} = B^i(0,1,0)\, e^{i(k_{sx}x+k_{sz}z-\omega t)} \tag{2.89}$$

SH wave

$$\psi^{i,\mathrm{SH}} = C^i(k_{sx},0,-k_{sz})\frac{c_s}{\omega}e^{i(k_{sx}x+k_{sz}z-\omega t)}, \tag{2.90}$$

where we have introduced the components of the wavevectors:

$$\mathbf{k}_p = (k_{px},0,k_{pz}) \tag{2.91}$$

and similarly for $\mathbf{k}_s$.

In general, there will be waves transmitted across the boundary, into a medium with different elastic constants (denoted by primes):
P wave

$$\phi^t = A^t e^{i(k'_{px}x+k'_{pz}z-\omega t)} \tag{2.92}$$

SV wave

$$\psi^{t,\mathrm{SV}} = B^t(0,1,0)\, e^{i(k'_{sx}x+k'_{sz}z-\omega t)} \tag{2.93}$$

SH wave

$$\psi^{t,\mathrm{SH}} = C^t(k'_{sx},0,-k'_{sz})\frac{c'_s}{\omega}e^{i(k'_{sx}x+k'_{sz}z-\omega t)}. \tag{2.94}$$

The corresponding reflected waves will have the x components of their wavevectors reversed, corresponding to waves moving away from the boundary, but we do not yet know the relationship between the incident and reflected wavevectors $\mathbf{k}$ and $\mathbf{k}''$:
P wave

$$\phi^r = A^r e^{i(-k''_{px}x+k''_{pz}z-\omega t)} \tag{2.95}$$

SV wave

$$\psi^{r,\mathrm{SV}} = B^r(0,1,0)\, e^{i(-k''_{sx}x+k''_{sz}z-\omega t)} \tag{2.96}$$

SH wave

$$\psi^{r,\mathrm{SH}} = C^r(-k''_{sx},0,-k''_{sz})\frac{c_s}{\omega}e^{i(-k''_{sx}x+k''_{sz}z-\omega t)}. \tag{2.97}$$

The total potentials on one side of the boundary are then

$$\phi = \phi^i+\phi^r \tag{2.98}$$

and

$$\psi = \psi^{i,\mathrm{SV}}+\psi^{i,\mathrm{SH}}+\psi^{r,\mathrm{SV}}+\psi^{r,\mathrm{SH}}, \tag{2.99}$$

and on the other side

$$\phi = \phi^t \tag{2.100}$$

and

$$\boldsymbol{\psi} = \boldsymbol{\psi}^{t,\mathrm{SV}} + \boldsymbol{\psi}^{t,\mathrm{SH}}. \tag{2.101}$$

Figure 4

Angles of incidence θ_p and θ_s, reflection θ''_p and θ''_s, and refraction θ'_p and θ'_s of elastic waves at a plane interface.

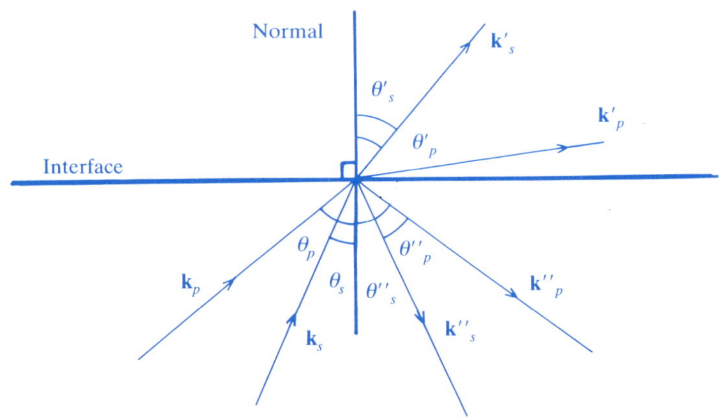

If we introduce the angles of incidence and refraction (Figure 4), we may write

$$\mathbf{k}_p = k_p(\cos\theta_p, 0, \sin\theta_p) \tag{2.102}$$

$$\mathbf{k}_s = k_s(\cos\theta_s, 0, \sin\theta_s) \tag{2.103}$$

$$\mathbf{k}'_p = k'_p(\cos\theta'_p, 0, \sin\theta'_p) \tag{2.104}$$

$$\mathbf{k}'_s = k'_s(\cos\theta'_s, 0, \sin\theta'_s) \tag{2.105}$$

$$\mathbf{k}''_p = k''_p(-\cos\theta''_p, 0, \sin\theta''_p) \tag{2.106}$$

$$\mathbf{k}''_s = k''_s(-\cos\theta''_s, 0, \sin\theta''_s). \tag{2.107}$$

In terms of components, then, the displacements and stresses from the P and SV waves are as follows:

$$u_x = \phi_{,x} - \psi_{y,z} \tag{2.108}$$

$$u_z = \phi_{,z} + \psi_{y,x} \tag{2.109}$$

$$\tau_{xz} = \mu(2\phi_{,xz} - \psi_{y,zz} + \psi_{y,xx}) \tag{2.110}$$

$$\tau_{xx} = \lambda(\phi_{,xx} + \phi_{,zz}) + 2\mu(\phi_{,zz} + \psi_{y,zx}), \tag{2.111}$$

and for the SH waves:

$$u_y = \psi_{x,z} - \psi_{z,x} \tag{2.112}$$

$$\tau_{xy} = \mu(\psi_{x,zz} - \psi_{z,xz}), \tag{2.113}$$

which shows clearly that the boundary will couple the P and SV waves together, but that the reflection and refraction of SH waves may be treated separately.

The unknown amplitudes of the reflected and transmitted waves are found by solving the conditions for continuity of stress and displacement at the interfaces. The resulting calculations are simpler if we take the limiting cases of rigid or free boundaries, with no transmitted waves, and we shall treat those cases before tackling the general case.

2.3.1 Reflection of SH waves by a boundary

The rigid boundary forces the displacements at $x=0$ to be zero, whereas a free boundary applies zero tractions at $x=0$. For the rigid boundary, the condition is therefore

$$u_y = 0 \tag{2.114}$$

that is,

$$\psi^{i,\mathrm{SH}}_{x,z} - \psi^{i,\mathrm{SH}}_{z,x} + \psi^{r,\mathrm{SH}}_{x,z} - \psi^{r,\mathrm{SH}}_{z,x} = 0 \tag{2.115}$$

which becomes, at $x=0$,

$$C^i e^{i(k_{sz}z-\omega t)} + C^r e^{i(k''_{sz}z-\omega t)} = 0. \tag{2.116}$$

If the boundary conditions are to be satisfied for arbitrary values of z, that is, for all positions on the boundary, the exponents must be the same. Thus k_{sz} and k''_{sz} must be equal. Referring back to equations (2.102) and (2.107) , this shows that θ_s and θ''_s must be equal: this is, of course, merely the expected result that the angle of incidence is equal to the angle of reflection. The same result will hold for all the cases of reflection and transmission that we treat below, and henceforth we drop the double prime notation and take it for granted that $\theta_s = \theta''_s$ and $\theta_p = \theta''_p$. We may also conclude that the amplitudes of the incident and reflected waves are equal in magnitude but opposite in sign: this corresponds to a change of phase of π radians on reflection.

For a surface which does not support tractions, which is a good approximation to the surface of a solid body in contact with air, we need to evaluate the stress tensor. There is only one non-zero component in this case, τ_{xy}:

$$\tau_{xy} = \mu(\psi_{x,zz} - \psi_{z,xz}). \tag{2.117}$$

For the incident wave this gives

$$\tau_{xy}^{i} = ik_{sx}\mu C^{i} e^{i(k_{sz}z-\omega t)} \tag{2.118}$$

and for the reflected wave

$$\tau_{xy}^{r} = -ik_{sx}\mu C^{r} e^{i(k_{sz}z-\omega t)}. \tag{2.119}$$

Here, then, we again find that the angles of incidence and reflection are equal, but in this case the incident and reflected amplitudes are equal – there is no change of phase.

2.3.2 Reflection of P and SV waves by a boundary

We have already seen, in equations (2.108) to (2.113), that a boundary can couple compression and shear waves together. This gives rise to mode conversion, which is so called because a pure compression wave incident on a boundary can give rise to a reflected shear wave as well as a reflected compression wave. The calculation of the reflection and transmission coefficients follows that for the SH waves, but is best expressed as a matrix equation because of the larger number of amplitude coefficients involved.

Taking, as before, the values at $x=0$, we obtain

$$u_x^{i} = ik_p \cos\theta_p A^{i} e^{ik_p z\sin\theta_p} - ik_s \sin\theta_s B^{i} e^{ik_s z\sin\theta_s} \tag{2.120}$$

$$u_x^{r} = -ik_p \cos\theta_p A^{r} e^{ik_p z\sin\theta_p} - ik_s \sin\theta_s B^{r} e^{ik_s z\sin\theta_s} \tag{2.121}$$

$$u_z^{i} = ik_p \sin\theta_p A^{i} e^{ik_p z\sin\theta_p} + ik_s \cos\theta_s B^{i} e^{ik_s z\sin\theta_s} \tag{2.122}$$

$$u_z^{r} = ik_p \sin\theta_p A^{r} e^{ik_p z\sin\theta_p} - ik_s \cos\theta_s B^{r} e^{ik_s z\sin\theta_s} \tag{2.123}$$

$$\begin{aligned}\tau_{xz}^{i} = \lambda(&-2k_p^2 \cos\theta_p \sin\theta_p A^{i} e^{ik_p z\sin\theta_p} \\ &+ k_s^2 \sin^2\theta_s B^{i} e^{ik_s z\sin\theta_s} - k_s^2 \cos^2\theta_s B^{i} e^{ik_s z\sin\theta_s})\end{aligned} \tag{2.124}$$

$$\begin{aligned}\tau_{xz}^{r} = \lambda(&2k_p^2 \cos\theta_p \sin\theta_p A^{r} e^{ik_p z\sin\theta_p} \\ &+ k_s^2 \sin^2\theta_s B^{r} e^{ik_s z\sin\theta_s} - k_s^2 \cos^2\theta_s B^{r} e^{ik_s z\sin\theta_s})\end{aligned} \tag{2.125}$$

$$\begin{aligned}\tau_{xx}^{i} &= -\lambda k_p^2 A^i e^{ik_p z\sin\theta_p} \\ &\quad +2\mu(-k_p^2\cos^2\theta_p A^i e^{ik_s z\sin\theta_s} - k_s^2\cos\theta_s\sin\theta_s B^i e^{ik_s z\sin\theta_s})\end{aligned} \tag{2.126}$$

$$\begin{aligned}\tau_{xx}^{r} &= -\lambda k_p^2 A^i e^{ik_p z\sin\theta_p} \\ &\quad +2\mu(-k_p^2\cos^2\theta_p A^i e^{ik_s z\sin\theta_s} + k_s^2\cos\theta_s\sin\theta_s B^i e^{ik_s z\sin\theta_s}).\end{aligned} \tag{2.127}$$

Clearly, if we are to satisfy the boundary conditions for all values of z, we must have

$$k_p\sin\theta_p = k_s\sin\theta_s \tag{2.128}$$

which may be written in terms of the wave speeds as

$$\sin\theta_p/\sin\theta_s = c_p/c_s. \tag{2.129}$$

This is the analogue for elastic waves of Snell's law in optics.

The exponential terms may now be factored out, and the matrix equations written down. If we specify a traction-free boundary, the displacements are unspecified, but the stresses result in the conditions:

$$\begin{aligned}&-2k_p^2\cos\theta_p\sin\theta_p A^i + k_s^2(\sin^2\theta_s - \cos^2\theta_s)B^i \\ &\qquad +2k_p^2\cos\theta_p\sin\theta_p A^r + k_s^2(\sin^2\theta_s - \cos^2\theta_s)B^r = 0\end{aligned} \tag{2.130}$$

$$\begin{aligned}&-(\lambda k_p^2 + 2\mu k_p^2\cos^2\theta_p)A^i - 2\mu k_s^2\cos\theta_s\sin\theta_s B^i \\ &\qquad +(\lambda k_p^2 + 2\mu k_p^2\cos^2\theta_p)A^r - 2\mu k_s^2\cos\theta_s\sin\theta_s B^r = 0\end{aligned} \tag{2.131}$$

or

$$\begin{aligned}&\begin{Bmatrix} -2k_p^2\cos\theta_p\sin\theta_p & k_s^2(\sin^2\theta_s - \cos^2\theta_s) \\ \lambda k_p^2 + 2\mu k_p^2\cos^2\theta_p & 2\mu k_s^2\cos\theta_s\sin\theta_s \end{Bmatrix}\begin{Bmatrix} A^i \\ B^i \end{Bmatrix} \\ &= \begin{Bmatrix} -2k_p^2\cos\theta_p\sin\theta_p & -k_s^2(\sin^2\theta_s - \cos^2\theta_s) \\ -\lambda k_p^2 - 2\mu k_p^2\cos^2\theta_p & 2\mu k_s^2\cos\theta_s\sin\theta_s \end{Bmatrix}\begin{Bmatrix} A^r \\ B^r \end{Bmatrix}.\end{aligned} \tag{2.132}$$

The equations may be simplified by noting that

$$\lambda + 2\mu\cos^2\theta_p = \rho(c_p^2 - 2c_s^2\sin^2\theta_p). \tag{2.133}$$

For incident compression waves, B^i is zero, and the solution is

$$R^{pp} \equiv \frac{A^r}{A^i} = \frac{\sin 2\theta_p\sin 2\theta_s - (c_p/c_s)^2\cos^2 2\theta_s}{\sin 2\theta_p\sin 2\theta_s + (c_p/c_s)^2\cos^2 2\theta_s} \tag{2.134}$$

$$R^{ps} \equiv \frac{B^r}{A^i} = \frac{2\sin 2\theta_p\cos 2\theta_s}{\sin 2\theta_p\sin 2\theta_s + (c_p/c_s)^2\cos^2 2\theta_s}. \tag{2.135}$$

Figure 5

Amplitudes of reflected compression and shear waves, for incident compression waves, for a range of Poisson ratios.

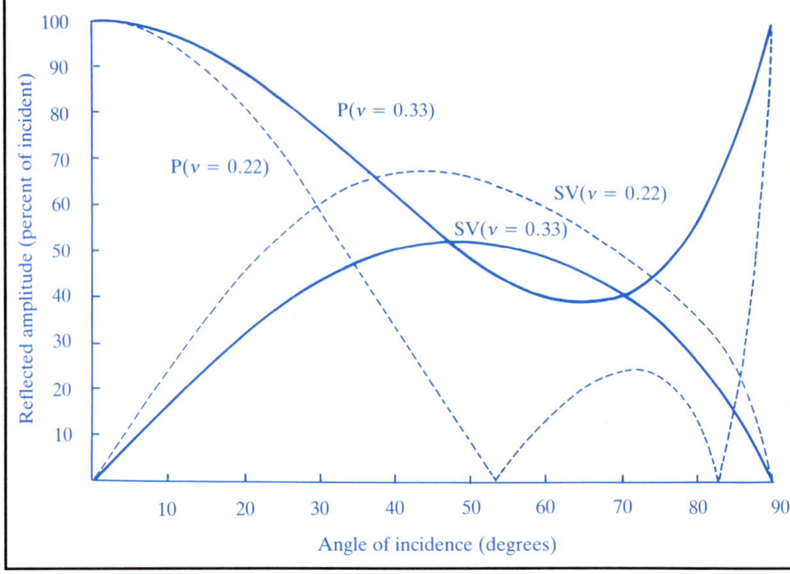

As Figure 5 shows, the mode conversion is zero for normal incidence and for glancing incidence. For angles of incidence in the region of 60–70° there is substantial loss of compression wave amplitude into shear waves.

The total energy in the waves is, however, conserved. The work done on a surface element dS by an elastic wave is given in equation (2.14) as the product of the surface traction $\mathbf{t}$ and the velocity of an element of the medium $\mathbf{u}_{,t}$. The surface element may be part of a hypothetical surface inside the body, when the traction $\mathbf{t}$ is $\boldsymbol{\tau}\cdot d\mathbf{S}$. Then, as in equation (2.70):

$$\mathbf{F} = \mathbf{u}_{,t}\cdot\boldsymbol{\tau}.$$

For a plane compression wave of the form given in equation (2.88) with propagation vector $\mathbf{k}_p$ from equation (2.102) the energy flux is then

$$\mathbf{F} = |A^i|^2 \frac{\rho}{c_p}\omega^4 \qquad (2.136)$$

directed parallel to $\mathbf{k}_p$. Similarly, for shear waves

$$\mathbf{F} = |B|^2 \frac{\rho}{c_s}\omega^4. \qquad (2.137)$$

The energy flux normal to the surface is therefore

$$F_i = |A^i|^2 \frac{\rho}{c_p} \omega^4 \cos\theta_p \tag{2.138}$$

in the incident compression waves, and a total of

$$F_r = |A^i R^{pp}|^2 \frac{\rho}{c_p} \omega^4 \cos\theta_p + |A^i R^{ps}|^2 \frac{\rho}{c_s} \omega^4 \cos\theta_s \tag{2.139}$$

in the reflected waves. After some algebra, the incident and reflected energies may be seen to be equal.

For incident shear waves, A^i is zero, and the solution is

$$R^{sp} \equiv \frac{A^r}{B^i} = \frac{-2(c_p/c_s)^2 \sin 2\theta_p \cos 2\theta_s}{\sin 2\theta_p \sin 2\theta_s + (c_p/c_s)^2 \cos^2 2\theta_s} \tag{2.140}$$

$$R^{ss} \equiv \frac{B^r}{B^i} = \frac{\sin 2\theta_p \sin 2\theta_s - (c_p/c_s)^2 \cos^2 2\theta_s}{\sin 2\theta_p \sin 2\theta_s + (c_p/c_s)^2 \cos^2 2\theta_s}. \tag{2.141}$$

Figure 6

Amplitudes of reflected compression and shear waves, for incident shear waves, for a range of Poisson ratios.

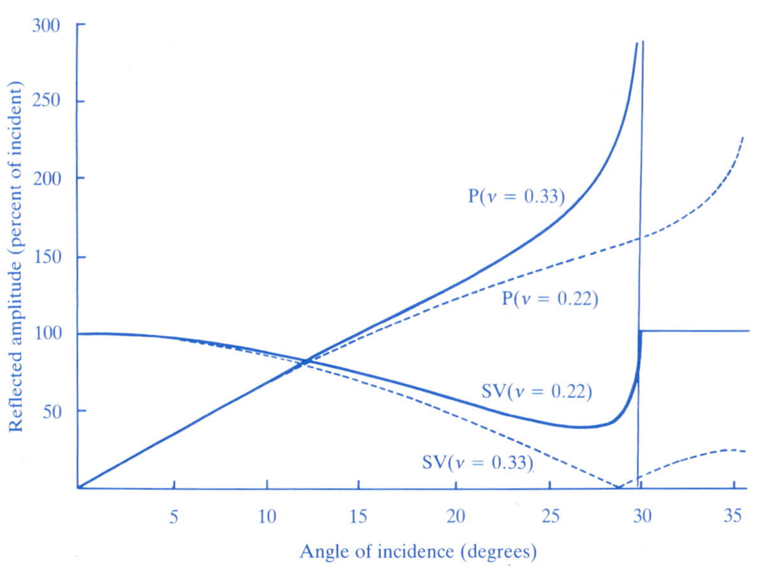

One important difference between this case and the case of incident compression waves is in the application of the Snell's law analogue. As c_p is always greater than c_s, there is a *critical angle* of incidence θ_c for shear waves, above which the value of $\sin\theta_p$ exceeds unity:

$$\theta_c = \sin^{-1}(c_s/c_p). \tag{2.142}$$

This means that above this angle there is no reflected compression wave, and Figure 6 shows this cutoff: typically the critical angle is about 30°. The other point to note is the marked dip in the reflected shear wave amplitude just below the critical angle. Such small reflectivities have important consequences in the use of elastic waves to detect crack-like defects.

The form of the compression wave for shear wave reflection beyond the critical angle deserves further comment. If we trace back beyond equations (2.102) – (2.103), we see that more fundamental quantities than $k_p\sin\theta_p$ and $k_p\cos\theta_p$ are k_{px} and k_{pz}, and that the boundary conditions require that

$$k_s\sin\theta_s = k_{pz}. \tag{2.143}$$

Now as

$$\frac{\omega^2}{c_p^2} = k_{px}^2 + k_{pz}^2 \tag{2.144}$$

it is possible for k_{pz} to take any value, to satisfy the boundary conditions, but this may be at the expense of an imaginary k_{px}. From equation (2.88), this means a compression wave with an amplitude which is exponentially increasing or decreasing with distance from the interface. Thus, even beyond the critical angle, there is a compression wave involved, but it is confined to a region close to the interface. Elastic energy is transported parallel to the interface by this wave.

2.3.3 Reflection and refraction at an interface

When elastic waves encounter an interface between two media, the boundary conditions require that displacements and stresses shall be continuous, rather than that they shall take specific values as in the previous section.

In general the boundary condition which is encountered is one of intimate contact, with normal and tangential components of traction and displacement continuous across the boundary. Slipping boundaries are sometimes encountered, in which the interface can transmit normal traction but not tangential traction. Then one requires continuity of normal stress and normal displacement and zero tangential stress on each side of the interface: there is no boundary condition on the tangential displacement, which may be discontinuous across the interface. Here we consider interfaces of the first type, and assume that only P and SV waves are involved: the geometry is the same as before. We then have incident, reflected and transmitted fields, and at an interface between two solids:

$$\tau^i_{xx} + \tau^r_{xx} = \tau^t_{xx} \quad (2.145)$$

$$\tau^i_{xz} + \tau^r_{xz} = \tau^t_{xz} \quad (2.146)$$

$$u^i_x + u^r_x = u^t_x \quad (2.147)$$

$$u^i_z + u^r_z = u^t_z. \quad (2.148)$$

As before, with incident P and SV wave amplitudes and four unknown amplitudes for the reflected P and SV and transmitted P and SV waves, a matrix equation may be set up and solved, using the four boundary condition equations (2.145) to (2.148). The explicit forms of the solutions may be found in standard texts (for example, Ewing, Jardetzky and Press 1957), and will not be reproduced here. For practical purposes, it is generally easier to set up and solve the equations using a numerical method on a computer: it is then easy to extend the code to treat layered systems. These are of interest when studying the coupling of transducers to test pieces through layers of grease or compliant rubbers.

Figure 7

Amplitudes of compression and shear waves transmitted into steel when compression waves are incident from water at various angles of incidence. For steel, c_p = 6.0 mm/μs, c_s = 3.3 mm/μs, ρ = 7.9 g/cm^3; for water c_p = 1.5 mm/μs, ρ = 1.0 g/cm^3.

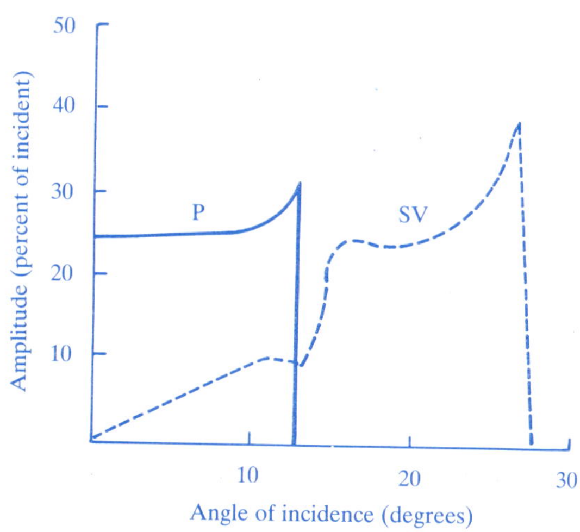

Considerable simplifications may be made when elastic waves are incident normal to an interface, so that no mode conversion takes place. It is then convenient to use the impedance concept, and to calculate the reflected and transmitted intensities using the standard formulae for scalar waves. The impedances are defined by equations (2.77) and (2.78) for compression and shear waves respectively. Then if the wave is incident from a medium of impedance Z_1 on an interface with a medium of impedance Z_2 the ratio of the reflected intensity F^r to the incident intensity F^i is given by

$$\frac{F^r}{F^i} = \left\{\frac{Z_2 - Z_1}{Z_2 + Z_1}\right\}^2, \tag{2.149}$$

and the transmitted intensity F^t is given by

$$\frac{F^t}{F^i} = \frac{4Z_2Z_1}{\{Z_2 + Z_1\}^2}. \tag{2.150}$$

In some applications of ultrasonic inspection the transducer is separated from the test specimen by a layer of water. In such cases it is important to know how much elastic energy will be transmitted into the specimen. Figure 7 shows the result of such a calculation for waves passing from water into steel. Although the water supports only compression waves, both compression and shear waves are generated in the steel. The angular variation of the transmission is important. For example, if a transducer in water generates a beam which is symmetrical about its axis, and this is incident at an angle onto steel with the aim of producing a shear wave in the metal, the beam in the steel will no longer be symmetrical, but will be enhanced in the regions of the beam corresponding to larger angles of incidence, that is, in the regions closer to the surface. This effect is distinct from the phenomenon of surface wave generation to be treated in a later section.

In Figure 7, as in all the figures in this section, the amplitudes of the waves are shown as percentages of the amplitude of the incident wave. To convert these amplitudes to energies they should be multiplied by the square of the frequency, squared, multiplied by the density and divided by the wave speed.

Once the wave has entered the steel, the reverse effect becomes important. We saw above that, for a free boundary, shear waves in steel would be totally internally reflected above the critical angle. Figure 8 shows that this is no longer true when the steel is immersed in a water bath. About 20 per cent of the energy is lost to the water on reflection above the critical angle. This is important if attempts are being made to cover a large range in a plate or a pipe wall by repeatedly reflecting elastic waves between the surfaces of the steel: in a submerged structure the range will be limited by loss into the water.

Figure 8

Amplitudes of compression and shear waves reflected when a shear wave in steel strikes a steel–water interface at various angles of incidence. For steel, c_p = 6.0 mm/μs, c_s = 3.3 mm/μs, ρ = 7.9 g/cm³; for water c_p = 1.5 mm/μs, ρ = 1.0 g/cm³.

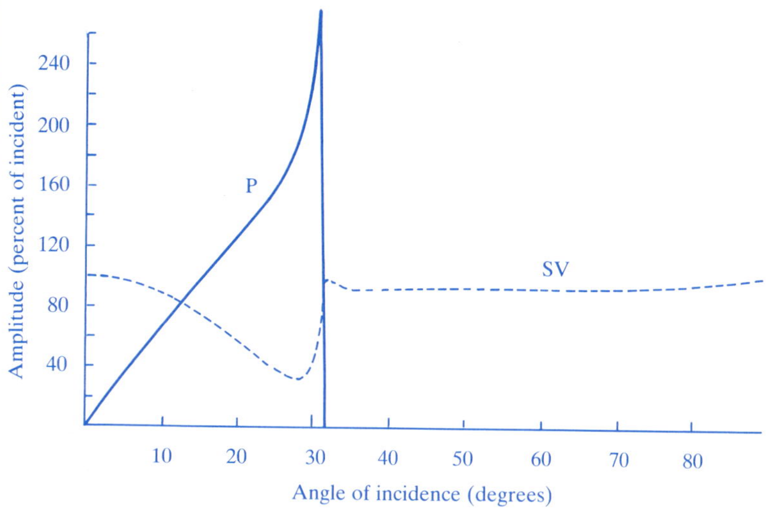

Similar losses occur in a material which carries a protective coating. Epoxy resins, and even pitch, behave like solids at ultrasonic frequencies. Figure 9 shows the reflection characteristics of a steel–epoxy interface. The same loss of energy in the region beyond the critical angle is observed.

In transferring elastic energy from one medium to another, the efficiency of transfer is controlled both by the sound speeds and the densities of the materials. Figure 10 exhibits this for an interface between steel and aluminium. The two metals have very similar wave speeds, but their densities are so different that a large fraction of the energy is reflected from the interface.

It is interesting to show the effects of thin layers of material on the transmission of elastic waves. A very wide crack in a wet environment might fill with water, and energy could then be transmitted across the crack through the water. This is probably not a suitable model for a narrow fatigue crack, where small regions of contact between faces might be expected, but it does model the simple slits which are often cut for laboratory experiments on 'cracks'.

Finally, it should be noted that refraction effects do not rely on the presence of a physical interface between different media. If the wave speeds change gradually within a body, then beams of light will be bent. This phenomenon is, of course, familiar in optics, where it gives rise to

Figure 9

Amplitudes of compression and shear waves reflected when a shear wave in steel strikes a steel–epoxy interface at various angles of incidence. For steel, $c_p = 6.0$ mm/μs, $c_s = 3.3$ mm/μs, $\rho = 7.9$ g/cm^3; for epoxy $c_p = 3.2$ mm/μs, $c_s = 1.6$ mm/μs, $\rho = 1.0$ g/cm^3.

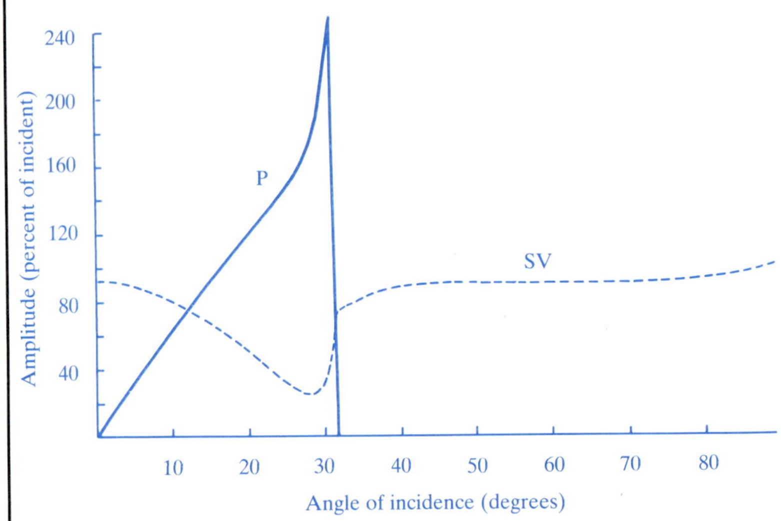

mirages. The ray tracing techniques of optics may be adapted (see Section 3.3) to elastic waves and used to analyse such cases. It is not necessary to have large changes in wave speeds: Smith, Richardson and van der Merwe (1987) show how the changes of 0.06 per cent in wave speed arising from residual stress in long shafts gave rise to 'blind spots' which could be misinterpreted as evidence of high attenuation.

2.4 Surface and guided waves

We have just seen that it is possible to generate a compression wave which is localized at an interface but sustained by an incident shear wave. This leads us to the more general question of interface waves, and we ask whether it is possible to have a wave confined to an interface but without an incident plane wave.

2.4.1 Rayleigh waves

Rayleigh (1885) studied the case of a traction-free surface: in effect he looked for solutions of equation (2.132) with no incident waves. Such solutions are called Rayleigh waves, or surface waves. There will be non-trivial solutions if the determinant of the coefficient matrix is zero, that is

Figure 10

Amplitudes of compression and shear waves reflected when a shear wave in steel strikes a steel–aluminium interface at various angles of incidence. For steel, $c_p = 6.0$ mm/μs, $c_s = 3.3$ mm/μs, $\rho = 7.9$ g/cm^3; for aluminium $c_p = 6.4$ mm/μs, $c_s = 3.2$ mm/μs, $\rho = 2.7$ g/cm^3.

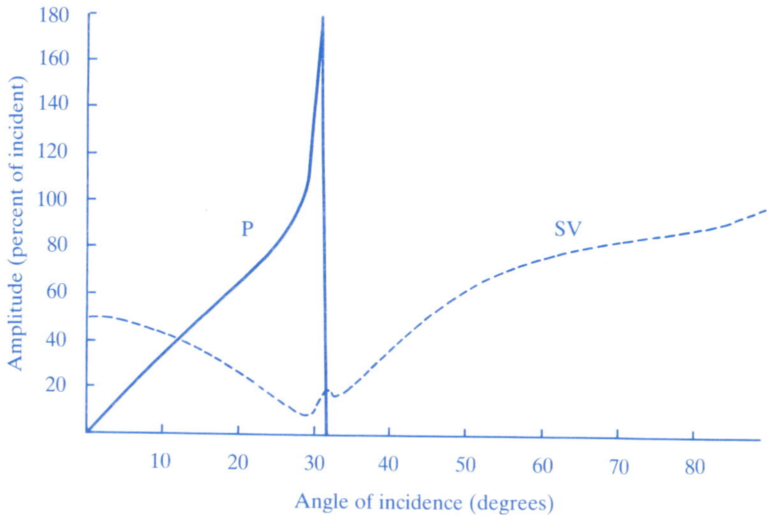

Figure 11

The amplitude of transmitted compression waves when a compression wave in steel strikes a water-filled layer at various angles of incidence. For steel, $c_p = 6.0$ mm/μs, $c_s = 3.3$ mm/μs, $\rho = 7.9$ g/cm^3; for water $c_p = 1.5$ mm/μs, $\rho = 1.0$ g/cm^3. The thickness of the slit is given as a multiple of the wavelength of the shear wave in steel.

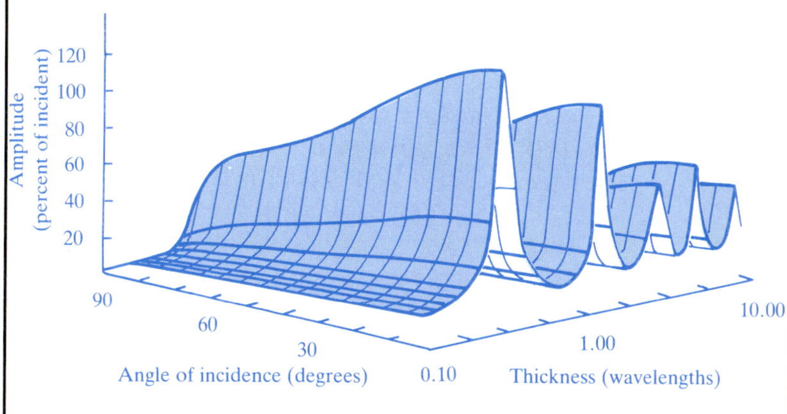

$$\begin{vmatrix} -2k_p^2\cos\theta_p\sin\theta_p & k_s^2(\sin^2\theta_s-\cos^2\theta_s) \\ k_p^2((c_p/c_s)^2-2\sin^2\theta_p) & 2k_s^2\cos\theta_s\sin\theta_s \end{vmatrix} = 0. \tag{2.151}$$

We must interpret the sines and cosines in terms of the components as

$$\cos\theta_p = k_x/k_p \tag{2.152}$$

$$\sin\theta_p = \sqrt{1-k_x^2/k_p^2} \tag{2.153}$$

$$\cos\theta_s = k_x/k_s \tag{2.154}$$

$$\sin\theta_s = \sqrt{1-k_x^2/k_s^2} \tag{2.155}$$

with k_x greater than both k_p and k_s to give an exponential decay with distance from the surface. If we introduce the Rayleigh wave velocity c_R, we can write the condition for the determinant to be zero in terms of velocities only as

$$(2-c_R^2/c_s^2)^2-4(1-c_R^2/c_p^2)^{\frac{1}{2}}(1-c_R^2/c_p^2)^{\frac{1}{2}} = 0. \tag{2.156}$$

It is clear that the requirement that $k_x>k_s$ requires $c_R<c_s$. The left-hand side of equation (2.156) changes sign as c_R increases from zero to c_s: for $c_R = \alpha c_s$ where α is small, it is $-2(1-c_s^2/c_p^2)\alpha^2$, and for $c_R = c_s$ it is unity. It is therefore clear that there is at least one solution for c_R. In other words, there is always at least one freely propagating surface wave on a free surface. Furthermore, as the frequency has cancelled out of the equation, the surface waves are nondispersive.

Analysis of the roots of equation (2.156) is made easier if we remove the square roots, and define

$$R \equiv (c_R/c_s)^2 \tag{2.157}$$

$$K \equiv (c_s/c_p)^2 \tag{2.158}$$

when

$$(2-R)^4 = 16(1-R)(1-KR) \tag{2.159}$$

or, defining

$$S \equiv R+\tfrac{8}{3}, \tag{2.160}$$

$$S^3+aS+b = 0 \tag{2.161}$$

$$a = (72-432K)/27 \tag{2.162}$$

$$b = (272-720K)/27. \tag{2.163}$$

We must remember that extraneous roots will be introduced by the squaring process. In fact, if we refer back to equation (2.132), we can see that the squaring process will introduce extra solutions which correspond to the zeros of the reflected compression wave amplitude for incident compression waves and of reflected shear wave amplitude for incident shear waves. The general rules for solutions of a cubic equation tell us that there will be three real solutions if

$$b^2/4+a^3/27 < 0, \tag{2.164}$$

and one real and two complex if

$$b^2/4+a^3/27 > 0: \tag{2.165}$$

that is, three real roots if

$$(17-45K)^2<(12K-2)^3 \tag{2.166}$$

and one real and two complex otherwise. The change in root pattern comes at K=0.321, that is for c_p<1.765c_s there are three real roots. Again, if we refer back to the reflection amplitudes in Figures 5 and 6 we can see that there is a change from the existence of zeros in the reflection coefficients to non-existence at about this ratio of wave speeds. More careful consideration of the three real roots shows that one solution lies in the region $0\leqslant c_R\leqslant c_s$, and the other two are larger than both c_s and c_p. Analysis of the original equation (2.156) using complex variable theory (for example, Achenbach 1973, page 189) shows that there is only one solution: the other two solutions introduced by squaring are spurious. The complex k plane must be cut to cope with the possible choices of phase in the square roots in equation (2.156), and the spurious solutions do not lie on the same Riemann sheet as the physical solution. A careful consideration of this point is given by Frisk, Dickey and Überall (1975), who show that a unique choice of phase is possible if the plane surface carrying the Rayleigh wave is considered as the limiting case of a cylinder of large radius. However, as Hudson (1980) points out, the fact that these extra solutions lie on different Riemann sheets does not mean that their effects are totally negligible. As we shall see later, the left-hand side of equation (2.156) often appears as the denominator inside an integral, so the zeros of the expression become poles of the integrand. Then the extra zeros may lie close enough to the integration contour to have significant effects: in fact the waves they give rise to are quite weak (Gilbert and Laster 1962).

A good approximation to the Rayleigh wave velocity is given by Viktorov (1967):

$$\frac{c_R}{c_s} = \frac{0.862+1.14\nu}{1+\nu}. \tag{2.167}$$

Typical values are given in the following table.

Rayleigh wave speeds		
ν	c_s/c_p	c_R/c_s
0.00	0.707	0.862
0.25	0.577	0.919
0.33	0.500	0.932
0.40	0.408	0.941
0.50	0.000	0.955

Figure 12

The displacements normal (u_z) and parallel (u_x) to the surface are plotted relative to the normal displacement at the surface, as functions of the ratio of the depth to the wavelength, for a Rayleigh wave on a material with Poisson's ratio of 0.25.

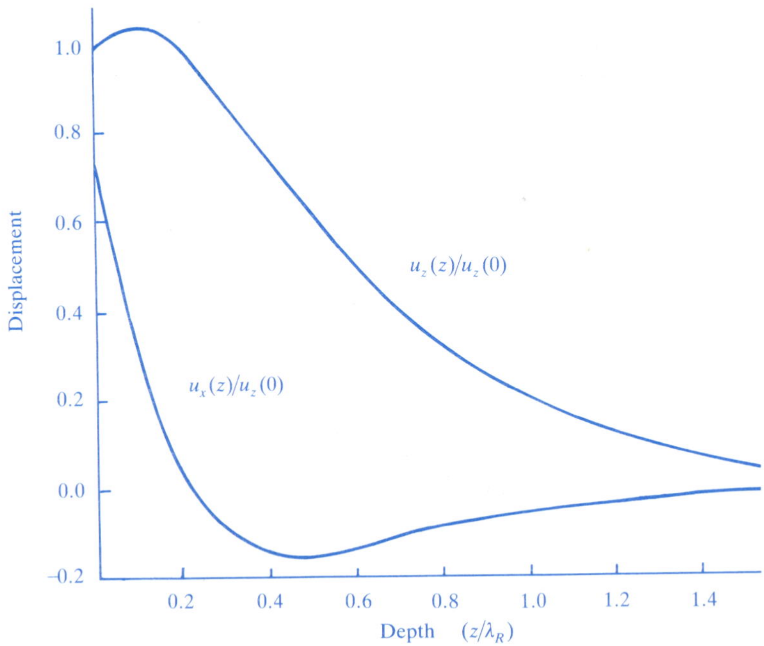

The exponential variation of the displacements in the Rayleigh wave with depth below the surface is shown in Figure 12. It is clear that the amplitudes are negligible more than about two wavelengths below the surface. However, as surface waves spread out over a two-dimensional surface instead of throughout a three-dimensional body their amplitudes as a function of distance from a small source decay only as $r^{-\frac{1}{2}}$ instead of as r^{-1}. This means that they can be important when surface-contacting transducers are used. The normal and tangential displacements at the surface are 90° out of phase, so the motion is elliptical. The characteristic 'rolling' motion of the strongest part of an earthquake is thus attributable to the fact that the signals which decay more slowly with distance from the disturbance are the surface waves, and they exhibit this elliptical motion.

Rayleigh waves can also be found on the surfaces of anisotropic materials. The existence and uniqueness of such waves has been studied by Barnett, Lothe, Nishioka and Asaro (1973), Chadwick and Smith (1977, 1982), Barnett and Lothe (1974, 1985), and Lothe and Barnett (1976). Barnett and Lothe (1985) use a method based on the impedance tensor to derive formulae which are particularly well suited to numerical computation.

2.4.2 Stoneley waves

In nondestructive testing, one often encounters situations where the test piece is in contact with another material. Obvious cases include submerged or fluid-containing structures, with a solid–fluid interface, and buried or coated structures, where the second medium is a solid which may have rather ill-defined characteristics. One then asks whether there are modes similar to Rayleigh waves on such interfaces. Love (1911) and Stoneley (1924) studied this problem: obviously it is related to the solution of the solid–fluid interface problem with no incident waves in the same way as the Rayleigh problem was related to the solution of equation (2.132).

For the solid–fluid case, there are two solutions, one complex and one real. As the density of the overlying fluid tends to zero, the complex solution tends to the Rayleigh wave velocity. It is complex because the Rayleigh wave is attenuated, and examination of the wavevector in the fluid shows that there is a real component normal to the surface. The Rayleigh wave on the surface of the solid is leaking into a bulk (compression) wave in the fluid. The real solution has a speed less than the wave speed in the fluid or the shear wave speed in the solid, whichever is the smaller, and has imaginary components of wavevector on both sides of the interface. This is an undamped interface wave, the Stoneley wave. If the density of the fluid is much less than that of the solid, which is the usual case in nondestructive testing applications, most of the energy of the Stoneley wave propagates through the fluid and most of that in the leaky Rayleigh wave through the solid, and the Stoneley wave may be neglected in most applications.

When the interface is between two solids Caignard (1962) has given a complete analysis. It turns out that a Stoneley wave occurs only for a restricted range of ratios of the densities and shear moduli of the two media. In particular, the shear wave velocities have to be quite similar for a Stoneley wave to exist, and this condition is unlikely to hold for a metallic structure buried in soil or coated with a resinous material. Even in the Earth's crust, where material properties might be expected to be similar enough to allow Stoneley wave propagation, seismologists have not observed such waves. Nevertheless, it has been suggested (Lee and Corbly 1977; Rokhlin, Hefets and Rosen 1980) that interface waves might prove useful for the examination of adhesively bonded joints. It is useful, then, to have an efficient procedure for calculating the properties of Stoneley waves when they exist, and such a method based on the impedance matrix has been developed by Barnett, Lothe, Gavazza and Musgrave (1985). Their method is suitable for anisotropic materials.

2.4.3 Lamb waves

In many practical applications of elastic waves the waves are propagated between two flat surfaces (or surfaces which have a large enough radius of curvature to be regarded as flat). From the discussion above it is clear that there will be complicated interconversions between compression and shear modes, but the energy of the wave will not be dissipated as it passes through the system. In this section we consider the properties of such 'guided waves' in plate-like structures.

The problem can be factored into components symmetric and antisymmetric about the centre plane of the plate. For the symmetric component, we take the potentials

$$\phi = A\cos\gamma_p z\, e^{i(kx-\omega t)} \tag{2.168}$$

$$\psi_y = B\sin\gamma_s z\, e^{i(kx-\omega t-\frac{\pi}{2})} \tag{2.169}$$

where we have introduced

$$\gamma_p = \sqrt{\frac{\omega^2}{c_p^2}-k^2} \tag{2.170}$$

$$\gamma_s = \sqrt{\frac{\omega^2}{c_s^2}-k^2}. \tag{2.171}$$

Note that the choice of phase in ϕ and ψ is made with the benefit of hindsight, and will result in real coefficients A and B.

The displacements are then

$$u_x = i\{Ak\cos\gamma_p z+B\gamma_s\cos\gamma_s z\}\, e^{i(kx-\omega t)} \tag{2.172}$$

$$u_z = \{-A\gamma_p\sin\gamma_p z+Bk\sin\gamma_s z\}\, e^{i(kx-\omega t)}. \tag{2.173}$$

This shows clearly the symmetry of the displacements,

$$u_x(-z) = u_x(z) \tag{2.174}$$

$$u_z(-z) = -u_z(z). \tag{2.175}$$

Note also that if A and B are real the x and z displacements will be in quadrature so that the particle motion will always be elliptical, with principal axes along x and z. At certain depths these ellipses collapse into linear motion.

To find the characteristic equations for Lamb waves we determine the surface stresses at the plate surfaces (at $z = \pm\frac{t}{2}$). The stresses (dropping the common exponential factor) are

$$\begin{aligned}
\tau_{zz}(\tfrac{t}{2}) &= (\lambda+2\mu)(-A\gamma_p^2\cos\gamma_p\tfrac{t}{2} + Bk\gamma_s\cos\gamma_s\tfrac{t}{2}) \\
&\quad -\lambda(Ak^2\cos\gamma_p\tfrac{t}{2} + Bk\gamma_s\cos\gamma_s\tfrac{t}{2}) \\
&= \mu(A(k^2-\gamma_s)^2)\cos\gamma_p\tfrac{t}{2} + 2Bk\gamma_s\sin\gamma_s\tfrac{t}{2}) \\
\tau_{xz}(\tfrac{t}{2}) &= i\mu(-Ak\gamma_p\sin\gamma_p\tfrac{t}{2} + Bk^2\sin\gamma_s\tfrac{t}{2}) \\
&\quad -i\mu(Ak\gamma_p\sin\gamma_p z + B\gamma_s^2\sin\gamma_s z) \\
&= -i\mu(2Ak\gamma_p\sin\gamma_p\tfrac{t}{2} + iB(\gamma_s^2-k^2)\sin\gamma_s\tfrac{t}{2}).
\end{aligned} \tag{2.176}$$

Setting the determinant of the coefficients of A and B to zero yields the characteristic equation for symmetrical Lamb waves

$$\frac{\tan\gamma_s\frac{t}{2}}{\tan\gamma_p\frac{t}{2}} = -\frac{4k^2\gamma_p\gamma_s}{(k^2-\gamma_s^2)^2}. \tag{2.177}$$

A similar analysis for the antisymmetric modes gives

$$\frac{\tan\gamma_s\frac{t}{2}}{\tan\gamma_p\frac{t}{2}} = -\frac{(k^2-\gamma_s^2)^2}{4k^2\gamma_p\gamma_s}. \tag{2.178}$$

It is convenient to express the characteristic equations in terms of the phase velocity v and the dimensionless frequency Ω, defining:

$$\upsilon \equiv \left(\frac{c_s}{v}\right)^2 \tag{2.179}$$

$$\kappa \equiv \left(\frac{c_s}{c_p}\right)^2 \tag{2.180}$$

$$\Omega \equiv \frac{\omega t}{2c_s}. \tag{2.181}$$

Then the characteristic equations may be written as

$$\frac{\tan(1-v)^{\frac{1}{2}}\Omega}{\tan(\kappa-v)^{\frac{1}{2}}\Omega} = \left\{\frac{4v(1-v)^{\frac{1}{2}}(\kappa-v)^{\frac{1}{2}}}{(2v-1)^2}\right\}^{\pm 1} \tag{2.182}$$

where the positive exponent is taken for symmetric modes, the negative for antisymmetric modes.

The solutions v of the above equation give the phase velocity of waves propagating without attenuation in plates. They consist of compression and shear waves propagating at angles θ_p and θ_s to the perpendicular to the plate surfaces, where

$$\sin\theta_p = \sqrt{\frac{v}{\kappa}} \tag{2.183}$$

$$\sin\theta_s = \sqrt{v}. \tag{2.184}$$

As in the case of a single free surface, we are not confined to real values of θ. There are regions where one or both of the compression and shear waves are evanescent, giving rise to the modified characteristic equations

$$\frac{\tan(1-v)^{\frac{1}{2}}\Omega}{\tanh(v-\kappa)^{\frac{1}{2}}\Omega} = \pm\left\{\frac{4v(1-v)^{\frac{1}{2}}(v-\kappa)^{\frac{1}{2}}}{(2v-1)^2}\right\}^{\pm 1} \tag{2.185}$$

in the region $\kappa<v<1$ and

$$\frac{\tanh(v-1)^{\frac{1}{2}}\Omega}{\tanh(v-\kappa)^{\frac{1}{2}}\Omega} = \left\{\frac{4v(v-1)^{\frac{1}{2}}(v-\kappa)^{\frac{1}{2}}}{(2v-1)^2}\right\}^{\pm 1} \tag{2.186}$$

when $v>1$. From this last case we can recover the Rayleigh wave equation in the high-frequency limit (when the left-hand side of the equation is unity). That is, when the frequency is so high that the wavelength is very small compared with the plate thickness one can propagate surface waves along each surface and their exponential decay with depth implies that they are independent. Both symmetric and antisymmetric combinations of Rayleigh waves are possible. In the same region a solution also occurs for small Ω, and this is the flexural mode of a thin plate.

For general wavelengths, the surfaces are coupled. Then only certain values of v are allowed because of the interference conditions between the waves (direct and mode-converted) reflected from the plate surfaces. This is made rather clearer by considering the limiting case of infinite phase velocity. This has every point along the length of the plate moving in phase: that is, it corresponds to a plane wave travelling perpendicular to the plate surfaces and being multiply reflected. As $v \to \infty$ so $\upsilon \to 0$ and the characteristic equation becomes

$$\frac{\tan\Omega}{\tan\sqrt{\kappa}\Omega} = \begin{cases} 0 & \text{for symmetric modes} \\ -\infty & \text{for antisymmetric modes.} \end{cases} \tag{2.187}$$

Thus, for symmetric modes the limiting cases occur for values of $\Omega = n\pi$ and $\sqrt{\kappa}\Omega = (n-\frac{1}{2})\pi$ for all integers n. If we write these in terms of the frequency $f = \frac{\omega}{2\pi}$, we find $ft = nc_s$, $(n-\frac{1}{2})c_p$. The limiting cases of the antisymmetric modes are interleaved with these, at $ft = (n-\frac{1}{2})c_s$, nc_p.

From each of the limiting cases springs a continuous series of roots (υ,Ω), maintaining the symmetry character of the $\upsilon = 0$ limit. The modes are labelled s or a according to their symmetry, and numbered from zero upwards according to the order in which they appear at $\upsilon = 0$. As υ is a function of Ω, the waves are dispersive: examples of the dispersion curves are given in Figure 13 .

In the limit of high frequency, apart from the mode described above which degenerates into two surface waves, the phase velocity tends to c_s and the displacements consist of layers of shear waves travelling parallel to the surface, separated by null planes of zero displacement. The number of null planes increases with the order of the mode, and is odd or even according to the symmetry.

The group velocity, which is also the speed at which energy is transported in the Lamb waves, is a function of frequency. It is related to the phase velocity

$$v = \frac{\omega}{k} \tag{2.188}$$

by

$$v_g = \frac{\partial\omega}{\partial k} \tag{2.189}$$

as shown above, and may be calculated from the relationship between v and Ω by

$$v_g = \frac{v}{1-\dfrac{\Omega}{v}\dfrac{\partial v}{\partial\Omega}}. \tag{2.190}$$

Curves of group velocity are shown in Figure 14 . Some aspects of the curves are worthy of comment. At $\Omega = 0$ the group velocities of s_n, a_n for $n>0$ are all zero: this corresponds to the plate thickness modes

Figure 13

Dispersion curves for the phase velocities of the first three symmetric and antisymmetric Lamb wave modes in a material with Poisson's ratio 0.29, given in terms of the reduced phase velocity $\frac{v}{c_s}$ and reduced frequency $\Omega \equiv \frac{\omega t}{2c_s}$

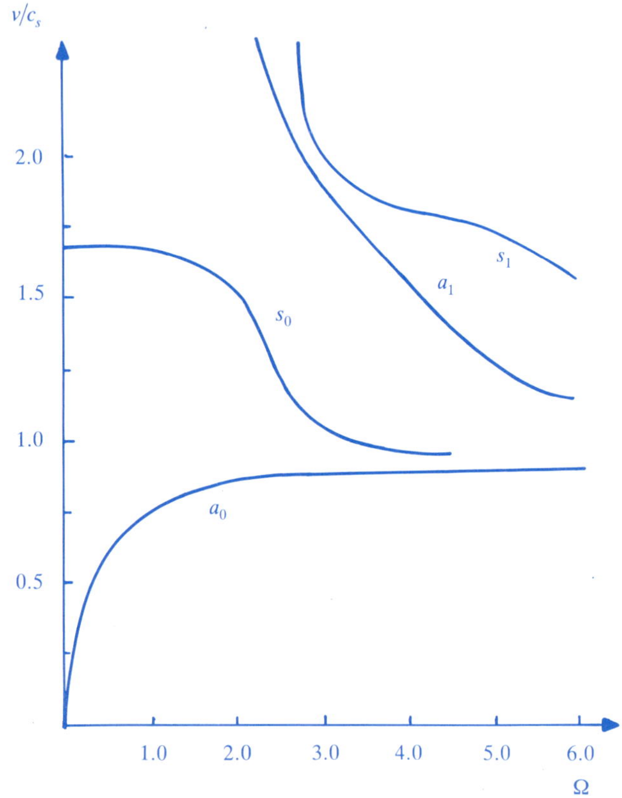

described above, for which there is obviously no energy flow along the plate. At the high frequency limit, the phase velocities approach c_s with ever decreasing slopes, so that the group velocities approach c_s from below. There are regions of negative group velocity, which simply means regions where the group and phase velocities are opposite. The dispersive properties of Lamb waves must satisfy the Kramers–Kronig relations, so that there must be regions of the spectrum where there is attenuation. For practical purposes, however, only the modes which propagate without attenuation are of interest.

For nondestructive testing applications, the most important feature of the dispersion curves is the existence of maxima and minima in the

Figure 14

Dispersion curves for the group velocities of the first three symmetric and antisymmetric Lamb wave modes in a material with Poisson's ratio 0.29, given in terms of the reduced group velocity $\frac{v_g}{c_s}$ and reduced frequency $\Omega = \frac{\omega t}{2c_s}$

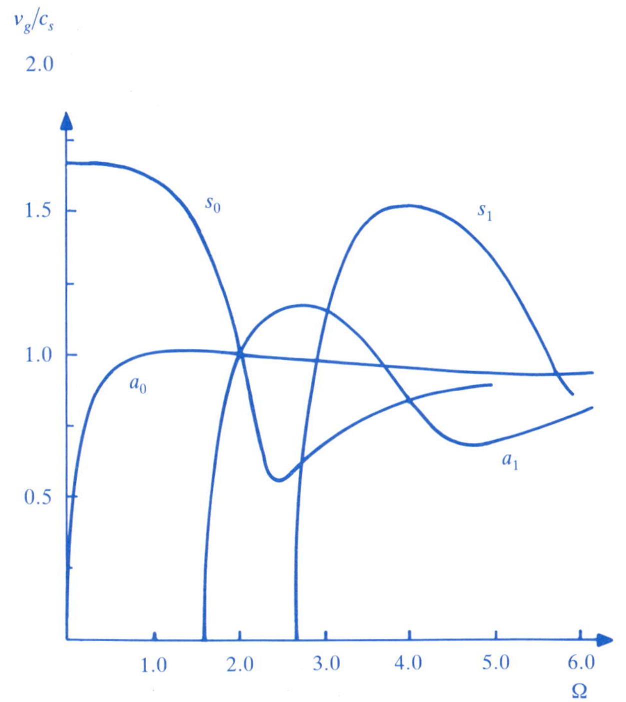

group velocity. These control the spreading of a finite pulse of elastic wave energy. Consider an initial pulse $f_0(t)$ applied to a line source in the plate at $x=0$:

$$f_0(t) = \int_{-\infty}^{\infty} \tilde{f}_0(\omega) e^{i\omega t} d\omega. \tag{2.191}$$

Suppose that each frequency propagates with a phase velocity $v(\omega)$; for the present we assume that only one wave mode is present. Then the disturbance seen at a point x will be

$$f(x,t) = \int_{-\infty}^{\infty} \tilde{f}_0(\omega) e^{i\omega(t - x/v(\omega))} d\omega. \tag{2.192}$$

The behaviour of $f(x,t)$ at large x may be derived using the method of stationary phase. This is a standard technique of mathematical physics, and the present case is similar to that discussed by Ewing, Jardetzky and Press (1957, page 143). The major contributions to $f(x,t)$ come from regions where the phase of the exponential is varying slowly. At such points

$$\frac{d}{d\omega}\omega(t-x/v(\omega)) = 0. \tag{2.193}$$

If ω_0 is a solution of equation (2.193) then

$$f(x,t) \sim \sqrt{2\pi} e^{i\chi} \tilde{f}_0(\omega_0) \left| x \frac{dv_g(\omega)}{d\omega} \right|^{-\frac{1}{2}} v_g(\omega) e^{i\omega_0(t-x/v(\omega_0))}. \tag{2.194}$$

Here χ is a phase which depends on the path taken for the stationary phase integral in the complex plane. The important point is the decay with distance as $x^{-\frac{1}{2}}$. Near a stationary point of v_g, the expression clearly breaks down. At such points (Morse and Feshbach 1953, page 631) $f(x,t) \sim x^{-\frac{1}{3}}$. This shows that at large distances the components which dominate the signal will be those that correspond to maxima or minima of the group velocity. If Lamb waves are to be used in nondestructive testing, then, it is sensible to select modes near such extrema, as these will allow fairly short pulses to be propagated without substantial losses in amplitude and time resolution arising from spreading.

When a Lamb wave propagates in a plate, it transfers energy parallel to the surface of the plate. The energy density in the wave is not spread uniformly throughout the plate: this is obvious when we recall that one limiting case of a Lamb wave consists of independent surface waves on the two faces of the plate. The energy density is the product of the stress and the particle velocity. It is readily seen that there is no energy flow in the y direction, because u_y, τ_{xy} and τ_{yz} are all zero. The energy flow in the z direction averages to zero over a whole cycle of the oscillating wave, as the products $\tau_{xz}u_x$ and $\tau_{zz}u_z$ both have a time dependence $\sin(2kx-2\omega t)$. Only the energy flow in the x direction has a non-zero average.

An example of the energy flow in a Lamb wave is shown in Figure 15 for an s_1 mode at $\Omega = 4.25$, with the magnitudes of the displacements as functions of depth in the plate. In this case, the power flow is concentrated near the plate surfaces. A knowledge of the regions of high energy density helps to select a wave mode which is sensitive to defects in particular regions of the plate. We shall return to the generation and scattering of Lamb waves in Chapter 4.

Figure 15

The displacements u_x and u_z, and the power flux along the plate, are shown for the s_1 Lamb wave with a reduced frequency $\Omega \equiv \frac{\omega t}{2c_s} = 4.25$.

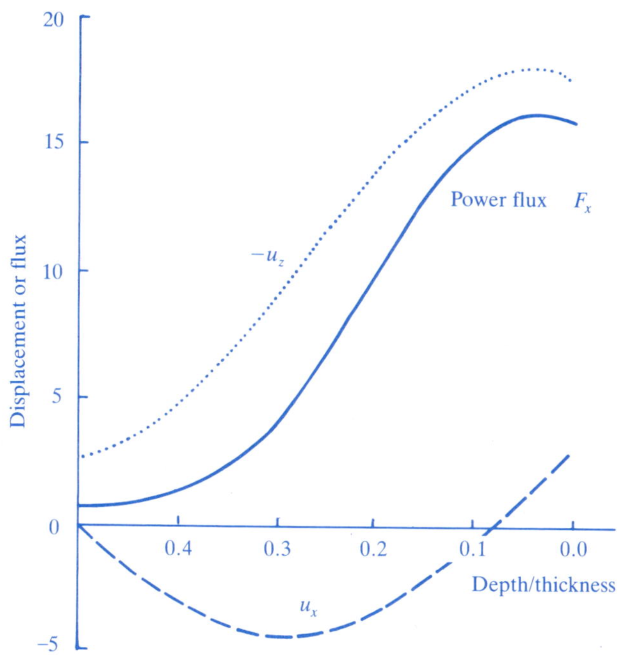

2.5 Bounded beams and pulses in solids

So far, we have treated plane waves of infinite extent and of a single frequency. In practice the elastic wavefields used will be of restricted width (because of the finite size of the transducers used to generate them) and of limited duration (which will mean that they contain a range of frequencies). We call these realistic wavefields bounded beams.

When beams have finite widths, some unexpected effects occur. Goos and Hänchen (1947) predicted that beams of light could be displaced laterally on reflection, and Schoch (1950) suggested that the same effect should occur in acoustics when a bounded beam is incident from a fluid onto a solid surface. This beam displacement effect has been discussed by several workers since then (Bertoni and Tamir 1973; Pitts and Mayer 1977; Pitts, Plona and Mayer 1977), and they predict and observe large changes in beam profile when a wave is incident from a fluid onto a solid near the Rayleigh angle. The Rayleigh angle is the angle of incidence in the fluid which matches a Rayleigh wave on the surface of the solid: in general, of course, this Rayleigh wave will be leaky (see above) and its speed (the real part of the complex wave speed) will differ from that of a free-surface Rayleigh wave, but we will assume that the elastic properties of the solid and the fluid are sufficiently different for us to use the free surface properties.

All that is required for a beam to be displaced is for the reflection coefficient to vary in amplitude, phase, or both, with the angle of incidence. This is quite easy to understand. We may represent a bounded beam as a sum of plane waves at different angles. Then in any angle-dependent reflection process different plane waves will have different coefficients, and the reflected beam will be reconstructed with a different profile from the input beam. The effects are usually quite small, as the variations of reflection coefficients with angle are usually quite small. However, as Figure 6 shows, the variations are much more rapid near critical angles, and there the bounded beam effects become important when one is comparing details of bounded beams on a scale comparable with the wavelength.

2.5.1 Infinite wave trains of finite width

Let us consider the reflection of a shear wave from a free surface above the critical angle. The reflection coefficient R^{ss} has been given in equation (2.141), and the phase η of that coefficient may be written as

$$\tan\tfrac{1}{2}\eta = -\frac{(c_p/c_s)\cos^2 2\theta}{2}\sin\theta\sin 2\theta((c_p/c_s)^2\sin^2\theta - 1)^{\frac{1}{2}}\,. \tag{2.195}$$

We express a shear wave of finite width as a superposition of plane shear waves with the same frequency dependence $e^{i\omega t}$ (which we shall not include specifically in the following equations) but different directions. So we take

$$\psi^i(\mathbf{x}) = \frac{1}{\sqrt{2\pi}}\int_{-\pi}^{\pi} B(\theta)e^{ik_s(\sin\theta x - \cos\theta y)}\,d\theta\,. \tag{2.196}$$

For a beam which is well directed along θ_0 we can choose $B(\theta)$ to be a symmetrical function of $\theta-\theta_0$ which is peaked at $\theta = \theta_0$. In this case the varying factor in the reflected beam is the phase shift $\eta(\theta)$, which is shown in Figure 16 . The reflected wave is

$$\psi^r(\mathbf{x}) = \frac{1}{\sqrt{2\pi}}\int_{-\pi}^{\pi} e^{i\eta(\theta)}B(\theta)e^{ik_s(\sin\theta x + \cos\theta y)}\,d\theta. \tag{2.197}$$

Now, if the beam is sufficiently wide that a relatively small angular range is included, and if the angle θ_0 is far enough from the stationary point at about 45°, the phase is approximately a linear function of θ. Then we write

$$\begin{aligned} R^{ss}(\theta) &= e^{i\eta(\theta)} \\ &= e^{i\eta(\theta_0)}e^{i(\theta-\theta_0)\eta'(\theta_0)}, \end{aligned} \tag{2.198}$$

Figure 16

The change of phase on reflection of an SV wave from a free surface as a function of angle of incidence above the critical angle for a material with Poisson's ratio $\frac{1}{3}$.

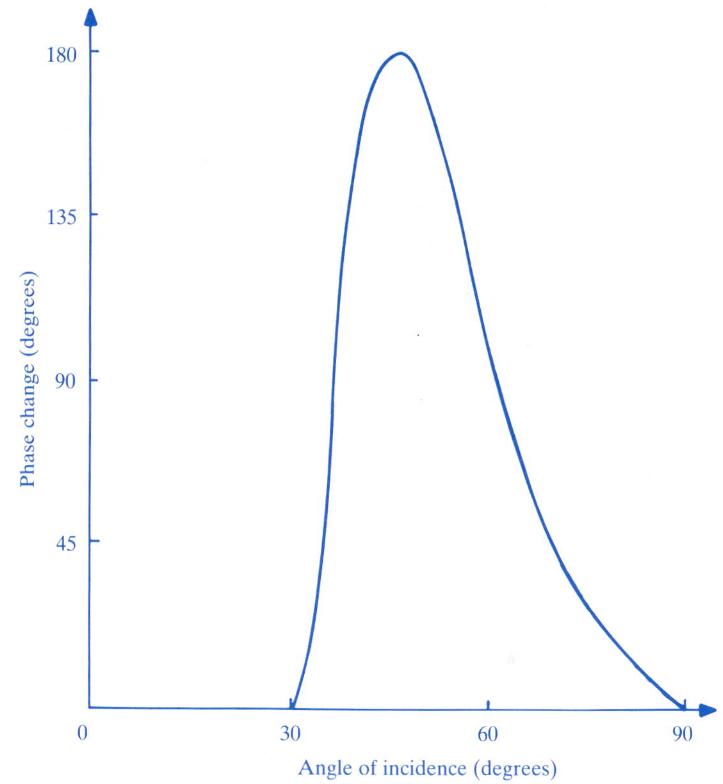

where η' is the first derivative of η with respect to θ.

Substituting and rearranging, we obtain

$$\begin{aligned}
\psi^r(\mathbf{x}) &= \frac{1}{\sqrt{2\pi}} e^{i\eta(\theta_0)} \int_{-\pi}^{\pi} e^{i(\theta-\theta_0)\eta'(\theta_0)} B(\theta) e^{ik_s(\sin\theta x+\cos\theta y)} d\theta \\
&\doteq \frac{1}{\sqrt{2\pi}} e^{i\eta(\theta_0)} \int_{-\pi}^{\pi} e^{i\sin(\theta-\theta_0)\eta'(\theta_0)} B(\theta) e^{ik_s(\sin\theta x+\cos\theta y)} d\theta \\
&= \frac{1}{\sqrt{2\pi}} e^{i\eta(\theta_0)} \int_{-\pi}^{\pi} e^{i(\sin\theta\cos\theta_0-\cos\theta\sin\theta_0)\eta'(\theta_0)} B(\theta) e^{ik_s(\sin\theta x+\cos\theta y)} d\theta \\
&= \frac{1}{\sqrt{2\pi}} e^{i\eta(\theta_0)} \int_{-\pi}^{\pi} B(\theta) e^{ik_s\left(\sin\theta\left(x+\frac{\eta'(\theta_0)}{k_s}\cos\theta_0\right)+\cos\theta\left(y-\frac{\eta'(\theta_0)}{k_s}\sin\theta_0\right)\right)} d\theta.
\end{aligned} \qquad (2.199)$$

Thus the reflected wave shows a shift of phase, and also a shift in the apparent position of the reflecting surface. The reflected wave on the surface $y = \eta'(\theta_0)\sin\theta_0/k_s$ is equal to the incident wave on the surface y=0 displaced by the distance $\eta'(\theta_0)\cos\theta_0/k_s$. Figure 16 shows that $\eta'(\theta_0)$ varies between 6 and 2 between 60° and 90°, so that the displacement is typically half a wavelength or less. This amount may be significant when attempts are being made to identify particular reflections. Depending on the angle of incidence, which controls the gradient, the displacement may be positive or negative.

2.5.2 Reflection and refraction of pulses

Considering again the case of total internal reflection, for a plane wave of infinite spatial extent but with a pulse shape $p(t)$, we may decompose the pulse into a set of plane waves, each of a single frequency, by taking the temporal Fourier transform. Then the reflection and refraction of each component wave may be treated by the theory developed in Section 2.3, and the resultants recombined to form the outgoing pulse. We begin by writing:

$$p(t) = \frac{1}{\sqrt{2\pi}}\int_{-\infty}^{\infty}\bar{p}(\omega)e^{-i\omega t}\,d\omega. \tag{2.200}$$

In this equation, we have to consider both positive and negative frequencies, ω. We have to consider carefully, however, the interpretation of the typical plane wave written as

$$\phi(\mathbf{r},t) = e^{i(\mathbf{k}\cdot\mathbf{r}-\omega t)}. \tag{2.201}$$

It is the relative sign of $\mathbf{k}\cdot\mathbf{r}$ and ωt which determines the direction of propagation. For negative frequencies one must alter not only the sign of the wavevector **k** to give a wave propagating in the right direction but also the sign of the imaginary part of $\cos\theta_p$ to give an exponentially decaying amplitude. Physically, this amounts to saying that the phase change η is consistently equivalent to a time shift in one direction. We may then write

$$R^{ss} = e^{-i\omega\eta(\theta)/|\omega|} \tag{2.202}$$

to emphasize the frequency dependence of the time shift $\eta(\theta)/|\omega|$. This will give rise to a reflected pulse which can most conveniently be treated (Ewing, Jardetzky and Press 1957, page 91) by transforming as follows:

$$\begin{aligned} p^r(t) &= \frac{1}{\sqrt{2\pi}}\int_{-\infty}^{\infty}\bar{p}(\omega)e^{-i\omega\eta(\theta)/|\omega|}e^{-i\omega t}\,d\omega \\ &= p(t)\cos\eta(\theta)-\bar{p}(t)\sin\eta(\theta) \end{aligned} \tag{2.203}$$

where

$$\bar{p}(t) = \int_{-\infty}^{\infty} d\omega \tilde{p}(\omega) e^{-i\omega t} \frac{1}{\sqrt{2\pi}} \frac{1}{i\pi} P \int_{-\infty}^{\infty} d\xi \frac{e^{-i\omega\xi}}{\xi}, \tag{2.204}$$

in which P denotes the principal part of an integral and

$$\frac{1}{i\pi} P \int_{-\infty}^{\infty} d\xi \frac{e^{-i\omega\xi}}{\xi} = \begin{cases} 1 & \omega<0 \\ 0 & \omega=0 \\ -1 & \omega>0 \end{cases} \tag{2.205}$$

is the part of the integral which is concerned with the different expressions for positive and negative frequencies.

Equation (2.203) shows that at the critical angle (η = 180°) and at normal incidence (η=0) the pulse is reflected without change of shape, but at other angles there may be a shape change.

As an example, consider the incident Gaussian pulse

$$\psi(\mathbf{x},t) = e^{-\alpha^2(\hat{\mathbf{k}}\cdot\mathbf{x}-c_s t)^2/c^2} \tag{2.206}$$

which has the Fourier transform

$$\Psi(\mathbf{x},\omega) = \frac{1}{\sqrt{2}\alpha} e^{i\omega\hat{\mathbf{k}}\cdot\mathbf{x}/c_s} e^{-\omega^2/4\alpha^2}. \tag{2.207}$$

The resulting reflected pulse has a different direction of propagation, $\hat{\mathbf{k}}'$, and the function $\bar{p}$ is

$$\begin{aligned} \bar{p}(\mathbf{x},t) &= \frac{1}{2\alpha\pi^{\frac{3}{2}} i} \int_{-\infty}^{\infty} d\xi \int_{-\infty}^{\infty} d\omega \frac{1}{\xi} e^{-i\omega(t+\xi-\hat{\mathbf{k}}'.\mathbf{x}/c_s)} e^{-\omega^2/4\alpha^2} \\ &= \frac{1}{i\pi} \int_{-\infty}^{\infty} d\xi \frac{1}{\xi} e^{-\alpha^2(t+\xi-\hat{\mathbf{k}}'.\mathbf{x}/c_s)^2} \\ &= \frac{2}{\sqrt{\pi}} e^{-\alpha^2(t-\hat{\mathbf{k}}'.\mathbf{x}/c_s)^2} \int_0^{\alpha(t-\hat{\mathbf{k}}'.\mathbf{x}/c_s)} e^{s^2} ds. \end{aligned} \tag{2.208}$$

The integral $e^{-x^2} \int_0^x e^{s^2} ds$, known as Dawson's integral, is available in standard tables (for example, Abramowitz and Stegun, 1965, page 319).

Clearly, for a phase shift η of 0 or 180° the overall pulse shape is unchanged. For intermediate angles, however, the pulse shape is altered as shown in Figure 17 . This phenomenon has also been described by Tolstoy (1973, page 83).

The change of shape can be further illustrated by the delta function pulse. Here if the incident pulse is

$$\psi^i(\mathbf{x},t) = \delta(\mathbf{k}\cdot\mathbf{x}-\omega t) \tag{2.209}$$

Figure 17

The change in shape of a gaussian pulse on reflection from a plane surface. η is the phase change in degrees.

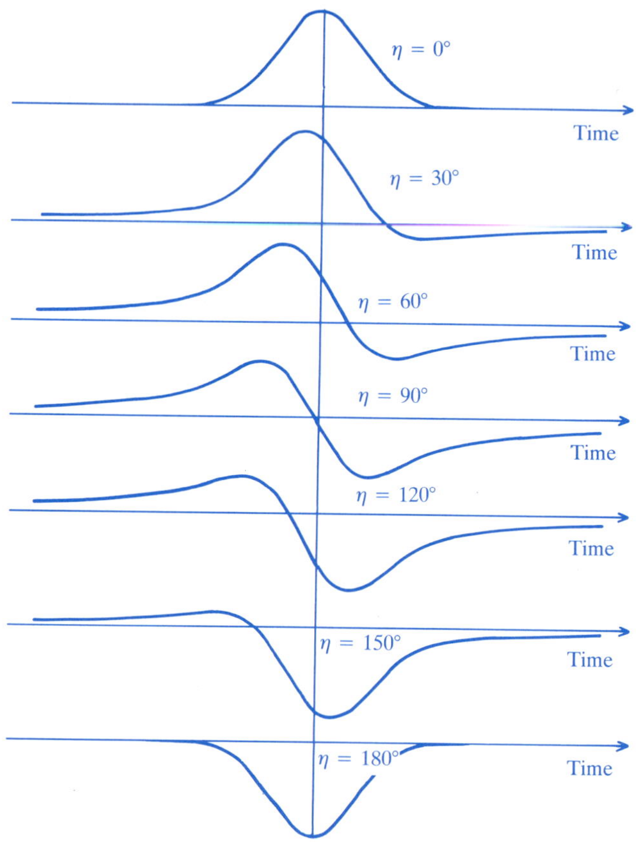

then the reflected pulse is

$$\psi^r(\mathbf{x},t) = \cos\eta\,\delta(\mathbf{k}\cdot\mathbf{x}-\omega t) - \frac{1}{\pi}\sin\eta\frac{1}{\mathbf{k}\cdot\mathbf{x}-\omega t}. \qquad (2.210)$$

This shows very clearly that the pulse has acquired both a tail and a precursor of infinite duration. This phenomenon is well known in seismology, where the precursor is known as the refracted arrival or the head wave.

At first sight this arrival of a signal *before* the expected start of the signal appears to violate the principle of causality. This is not so: there is a physical explanation for the early arrival, as described by

Friedlander (1948). Even when a shear wave is totally internally reflected there is a compression wave at the surface (as described above), and this can propagate along the surface from the 'edge' of the infinite wavefront, to arrive before the shear wave front. A similar effect occurs in optics: here total internal reflection occurs when the trace velocity (the velocity parallel to the surface) of a wave in an optically dense medium is less than the velocity of an evanescent wave along the interface in the less dense medium. One would expect no such effect for an acoustic wave incident on a free or rigid surface, where only one wave speed is involved, and analysis confirms that in that case the phase shift is 0 or 180°. For elastic waves at a free surface, changes of phase other than 0 or 180° occur only when a shear wave of SV polarization is incident beyond the critical angle.

In any practical application, the incident wavefield will be of finite width. As shown above, this implies that it is a superposition of plane waves at a range of angles. When the integration over angles is performed, the early arrivals cancel, giving an output pulse which is causal. A similar cancellation is discussed by Heyman and Felsen (1985) in the context of Chapman's (1978) scheme for calculating synthetic seismic traces. They show that the integral over angles may be split into a visible range, in which the wavevector is real, and an evanescent part with complex wavevector. Both the visible and evanescent parts contain terms which are non-zero in the non-causal time range, but they exactly cancel. The contribution of the evanescent part in the causal range is usually small compared with the visible part, so the integral over the angles which give real wavevectors gives quite a good approximation to the waveform if the non-causal early arrivals are removed.

2.6 Generation of elastic waves

The generation of elastic waves in a medium by the application of stress at the surface is crucial to all nondestructive testing applications: it represents the way in which the ultrasonic waves are inserted into a test-piece. The process is complicated by the structure of the elasticity tensor, so that when a normal force is applied to a surface a mixture of longitudinal and shear waves will be generated.

The simplest problem to investigate is the pattern of elastic waves generated by time-harmonic stresses. Problems of this sort were first solved by Lamb (1904) who also considered buried sources. Problems of the latter kind are important in seismology, where the sources of earthquakes are frequently buried in the Earth's crust. They are also of interest in nondestructive testing applications, especially in acoustic emission experiments where the source of the disturbance is a defect in the material. Acoustic emission has grown rapidly from a laboratory study by Kaiser (1953) through early application (Green, Lockman and Steel 1964) to become an established technique (Scruby and Wadley

1983), largely as a result of a firm theoretical understanding of its foundations. Lamb's theory has been reworked by many workers, partly to cast it into more elegant mathematical forms and partly to allow quantitative results to be extracted.

In the following section we shall describe the waves generated in a semi-infinite solid by a sinusoidally time-varying surface stress applied over a small region. This will show the important general features of the angular dependence of the waves generated.

2.6.1 Waves from time-harmonic surface stresses

The early work on the excitation of elastic waves by a harmonic load at the surface was undertaken by Lamb (1904), but we base the present summary on the papers by Miller and Pursey (1954, 1955). They analysed four cases: strip loads with normal and tangential stresses, and disk loads with normal and torsional stresses. We shall reproduce the main points of their argument for the strip normal stress. Consider a half-space $z<0$, subject to a unit normal stress on a strip $|x|<a$. If the stress is harmonic with angular frequency ω, and the problem is independent of y, the boundary conditions on $z=0$ are

$$\tau_{zz}(x,0,t) = \begin{cases} e^{i\omega t} & |x|<a \\ 0 & |x|>a \end{cases} \tag{2.211}$$

$$\tau_{xz}(x,0,t) = 0. \tag{2.212}$$

The solution may be expressed in terms of potentials for the compression and shear waves, ϕ and ψ, each of which satisfies a wave equation, of the form

$$\nabla^2\phi + k_p^2\phi = 0 \tag{2.213}$$

for the compression wave and

$$\nabla^2\psi + k_s^2\psi = 0 \tag{2.214}$$

for the shear wave. If we apply a spatial Fourier transform on the x variable,

$$\bar{f}(k) \equiv \frac{1}{\sqrt{2\pi}}\int_{-\infty}^{\infty} f(x)e^{-ikx}\,dx \tag{2.215}$$

with the inverse transform

$$f(x) \equiv \frac{1}{\sqrt{2\pi}}\int_{-\infty}^{\infty} \bar{f}(k)e^{ikx}\,dk\ , \tag{2.216}$$

then the wave equations become

$$\bar{\phi}_{,zz} - (k^2 - k_p^2)\bar{\phi} = 0 \tag{2.217}$$

and

$$\tilde{\psi}_{,zz} - (k^2 - k_s^2)\tilde{\psi} = 0, \tag{2.218}$$

with the solutions

$$\tilde{\phi} = A(k)e^{-\sqrt{k^2-k_p^2}z} \tag{2.219}$$

and

$$\tilde{\psi} = B(k)e^{-\sqrt{k^2-k_s^2}z}. \tag{2.220}$$

The relations between the stresses and the transformed potentials are then

$$\tilde{\tau}_{zz} = \lambda\tilde{\phi}_{,zz} - \lambda k^2\tilde{\phi} + 2\mu\tilde{\phi}_{,zz} + 2\mu ik\tilde{\psi}_{,z} \tag{2.221}$$

$$\tilde{\tau}_{xz} = \mu\left[2ik\tilde{\phi}_{,z} - k^2\tilde{\psi} - \tilde{\psi}_{,zz}\right]. \tag{2.222}$$

Similarly, the transformed boundary conditions are

$$\tilde{\tau}_{zz} = \frac{2\sin ka}{k} \tag{2.223}$$

$$\tilde{\tau}_{xz} = 0. \tag{2.224}$$

Substituting the explicit forms of the solutions, equations (2.219) and (2.220), into the expressions for the stresses in equations (2.221) and (2.222), and equating them to the boundary conditions in equations (2.223) and (2.224) at $z=0$ gives a pair of simultaneous equations for A and B:

$$\frac{2\sin ka}{\mu k} = -(2k^2 - k_s^2)A + 2ik(k^2 - k_s^2)^{\frac{1}{2}}B \tag{2.225}$$

$$0 = -2ik(k^2 - k_p^2)^{\frac{1}{2}}A - (2k^2 - k_s^2)B \tag{2.226}$$

which have the solutions

$$A = -\frac{2\sin ka}{\mu k}\,\frac{2k^2 - k_s^2}{F(k)} \tag{2.227}$$

$$B = \frac{2\sin ka}{\mu k}\,\frac{2ik(k^2 - k_p^2)^{\frac{1}{2}}}{F(k)} \tag{2.228}$$

where

$$F(k) \equiv (2k^2 - k_s^2)^2 - 4k^2(k^2 - k_p^2)^{\frac{1}{2}}(k^2 - k_s^2)^{\frac{1}{2}}. \tag{2.229}$$

It should be noted that $F(k)$ is the expression encountered before, in (2.156), with zeros which correspond to surface waves and to zeros of the compression–compression and shear–shear reflection coefficients of a free surface.

Substituting these values of A and B and evaluating the displacements, the inversion of the Fourier transforms requires the following integrals to be evaluated:

$$u_x(x,z) = \frac{1}{i\mu\pi}\int_{-\infty}^{\infty}\frac{\sin ka}{F(k)}\Big[2(k^2-k_p^2)^{\frac{1}{2}}(k^2-k_s^2)^{\frac{1}{2}}e^{-(k^2-k_s^2)^{\frac{1}{2}}z} + (k_s^2-2k^2)e^{-(k^2-k_p^2)^{\frac{1}{2}}z}\Big]e^{ikx}\,dk \tag{2.230}$$

$$u_z(x,z) = \frac{1}{\mu\pi}\int_{-\infty}^{\infty}\frac{(k^2-k_p^2)^{\frac{1}{2}}\sin ka}{kF(k)}\Big[2k^2(k^2-k_s^2)^{\frac{1}{2}}e^{-(k^2-k_s^2)^{\frac{1}{2}}z} + (k_s^2-2k^2)e^{-(k^2-k_p^2)^{\frac{1}{2}}z}\Big]e^{ikx}\,dk. \tag{2.231}$$

In the limit of a line source, sinka in these integrals should be replaced by ka.

To progress further we need to consider the properties of the integrands in the complex plane. It is clear that, in principle, the integrals may be performed by closing the integration contour in the upper half of the complex k–plane for positive x, in the lower half for negative x. However, the integrands contain square roots, leading to branch points at $k = \pm k_p$, $\pm k_s$, and in order to make the integral single valued it is necessary to keep to the correct branch.

Figure 18

The branch points, branch cuts, and poles of the integrand involved in the calculation of the wavefield arising from a surface load. The details of the branch cut contours Γ_2 and Γ_3 are given in the text.

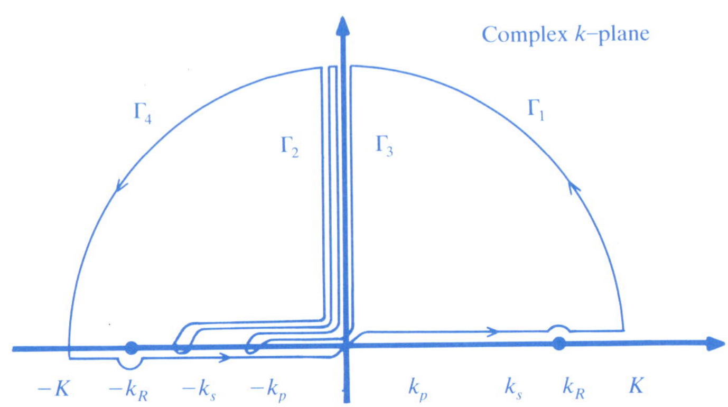

The choice of branch is made by noting the behaviour of the exponentials in y: we must always have $\mathcal{R}(k^2-k_p^2)^{\frac{1}{2}}$ and $\mathcal{R}(k^2-k_s^2)^{\frac{1}{2}}$ positive if the displacements are to be bounded throughout the half-space. Similarly the radiation condition tells us how the contour should be indented around the Rayleigh wavevector $\pm k_R$, the roots of $F(k)$. The resulting contour is shown in Figure 18. The contour integral is

$$\begin{aligned}\oint G(k) &= \int_{\Gamma_1} G(k) + \int_{\Gamma_2} G(k) + \int_{\Gamma_3} G(k) + \int_{\Gamma_4} G(k) + \int_{-K}^{K} G(k) \\ &= -2\pi i \Sigma[\text{Residues } of\ G(k)]\end{aligned} \tag{2.232}$$

in which the contributions from the segments at infinity, Γ_1 and Γ_4, vanish, and the branch cut paths are

$$\Gamma_2 = (0,iK) \rightarrow (0,0) \rightarrow (0,-k_p) \rightarrow (0,0) \rightarrow (0,iK) \tag{2.233}$$

and

$$\Gamma_3 = (0,iK) \rightarrow (0,0) \rightarrow (0,-k_s) \rightarrow (0,0) \rightarrow (0,iK). \tag{2.234}$$

The evaluation of the integrals along these branch cuts is the major task in the integration, and analytic results can only be obtained in the limit of large x and z.

The field at large distances may be evaluated by the method of steepest descents. This technique is covered in standard texts on mathematical physics, and only a brief summary will be given here. The typical integral which we have to evaluate has the form

$$I = \int_{-\infty}^{\infty} f(k) e^{ikx-(k^2-q^2)^{1/2}z}\, dk \tag{2.235}$$

where q takes the values k_p or k_s. If we introduce polar coordinates, with an angle θ measured from the z-axis,

$$x = R\sin\theta \qquad\qquad z = R\cos\theta \tag{2.236}$$

then

$$I = \int_{-\infty}^{\infty} f(k) e^{R\chi(k)}\, dk \tag{2.237}$$

where

$$\chi(k) = ik\sin\theta - (k^2-q^2)^{\frac{1}{2}}\cos\theta. \tag{2.238}$$

The technique now is to find the saddle points of $\mathcal{R}(\chi)$ and to deform the integration contour so as to pass through the saddle points on a steepest descents path. In this case there are two saddle points, where $d\chi(k)/dk = 0$, at

$$k_0 = \pm q\sin\theta, \tag{2.239}$$

but only the one at $k_0 = -q\sin\theta$ lies inside the present integration contour, shown in Figure 18. At this saddle point we have

$$\chi(k_0) = -iq \qquad \chi''(k_0) = i\sec^2\theta \tag{2.240}$$

and we expand the exponent about the saddle point as

$$\chi(k) = \chi(k_0)+(k-k_0)^2\chi''(k_0). \tag{2.241}$$

To convert the exponential term to the form e^{-u^2} we require

$$k-k_0 = \pm\sqrt{q}u\cos\theta e^{i\pi/4} \tag{2.242}$$

where the sign is determined by examining the direction at which the contour should cross the real k axis: the positive sign gives a path crossing the axis at 45 degrees, the negative sign a path at 135 degrees. For further details of the integration contour, the original papers by Miller and Pursey should be consulted.

Finally, we must evaluate the residue at the point k_R, which contains the term

$$e^{-R(k_R^2-q^2)^{\frac{1}{2}}\cos\theta} \tag{2.243}$$

which will be negligible except when $\theta = \pi/2$. This confirms that this is the surface wave contribution. Overall, then, the asymptotic form of the integral is

$$I\sim\sqrt{\frac{2\pi q}{R}}\,e^{i(\pi/4-qR)}\cos\theta f(-q\sin\theta). \tag{2.244}$$

Using equation (2.244) in the integrals for the displacement components (2.230) and (2.231) gives the asymptotic forms of the displacements. For simplicity we define the radial and tangential components u_R and u_θ by

$$u_R = u_z\cos\theta+u_x\sin\theta \tag{2.245}$$

$$u_\theta = -u_z\sin\theta+u_x\cos\theta \tag{2.246}$$

which results, for a line source, in

$$u_R = \frac{1}{\mu}\sqrt{\frac{2}{\pi k_p R}}\,e^{i(\frac{3}{4}\pi-k_pR)}\,\frac{\cos\theta(k_s^2-2k^2\sin^2\theta)}{F(k_p\sin\theta)} \tag{2.247}$$

$$u_\theta = \frac{1}{\mu}\sqrt{\frac{2}{\pi k_s R}}\,e^{i(\frac{5}{4}\pi-k_sR)}\,\frac{\sin2\theta(k_s^2\sin^2\theta-k_p^2)}{F(k_s\sin\theta)}. \tag{2.248}$$

In the far field, we may equate u_R with the compression waves and u_θ with the shear waves: the variations with angle are plotted in Figure 19. The dip in the shear wave amplitude at the critical angle is of interest. This is caused by the excitation of the head wave, which is a plane wave

Figure 19

The wave amplitudes of shear (left) and compression (right) modes as functions of angle from a line force applied normal to the surface.

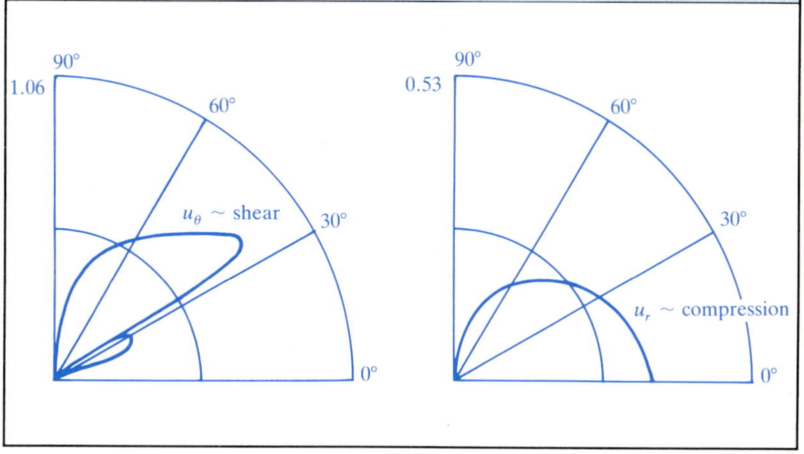

travelling at the critical angle and exactly cancelling the direct shear wave at that angle. We shall return to the head wave shortly.

By considering the case of $z=0$ separately, the surface wave amplitudes may be deduced. More than half the energy is propagated as surface waves (the precise fraction depends on Poisson's ratio). As surface waves spread in two dimensions whereas bulk waves spread in three dimensions, the surface waves decrease more slowly with distance than do bulk waves. This means that, in a nondestructive testing system, it is often necessary to have some means of damping surface waves if small surface blemishes are not to overwhelm the reflected signals from more important defects.

The expressions given above for u_R and u_θ are zero at the surface $z=0$, where $\theta = \frac{1}{2}\pi$. This far-field result obscures some important physics. If we consider the fields on the surface, the steepest descents method is no longer applicable, as the typical integral has the form

$$I = \int_{-\infty}^{\infty} f(k) e^{ik} dk \qquad (2.249)$$

and the appropriate contour for the integral is parallel to the imaginary k axis, which consists of circuits of the two branch points and the Rayleigh pole. The resultant displacements (for example, Hudson 1980) vary as $(ka)^{-\frac{3}{2}}$, and therefore decay with distance from the source faster than do the bulk waves. These are, then, modes of compression and shear type which propagate parallel to the free surface, but are attenuated. This amplitude decay occurs as they radiate true bulk waves continuously. Russian workers (Basatskaya and Ermolov 1980; Yuozonene 1980) have studied these subsurface modes for nondestructive testing applications. They are suitable for examining surface layers of materials such as plastics which have low sound velocities, in which true Rayleigh waves cannot easily be excited because of the difficulty of matching the impedance when inserting the waves from a second medium. It is important to realize, however, that they are not new modes. In particular, they are not the exponentially damped modes which arise as the spurious solutions of the Rayleigh wave problem, equation (2.156), discussed above.

2.7 The wave equation: integral form

The wave equation may be recast as an integral equation instead of a differential equation. This is a purely formal re-expression of the same mathematics, but it has some advantages when approximate methods of solution are sought. The basic idea is that if one knows the response of an elastic material to local forces with specified time dependencies, the principle of superposition tells us that we can build models of more complicated cases by combining these elementary solutions. The elementary forces are usually taken to be single-frequency simple harmonic waves, step functions, or delta functions. The responses to unit local forces define the Green tensor.

2.7.1 Green functions

For steady-state excitation with angular frequency ω, the integral equations have been derived by, among others, Kupradze (1963), de Hoop (1958) and Pao and Varatharajulu (1976). Morse and Feshbach (1953) also give the results.

A fairly straightforward derivation is by means of the Rayleigh–Betti reciprocity relation – in itself an important result. The result derived by Betti (1872) is a special case of the more general theorem by Rayleigh (1873). If there are two sets of displacements **u** and **u**$'$, corresponding to body forces **f** and **f**$'$ respectively, then the total of the work done by the first set of forces acting on the second set of displacements is equal to the work done by the second set of forces acting on the first set of displacements. From the divergence theorem, we may immediately write down

$$\begin{aligned}\int_V[\nabla_i(c_{ijkl}\nabla_l u_k)u'_j-\nabla_i(c_{ijkl}\nabla_l u'_k)u_j]dV\\ = \int_S[c_{ijkl}\nabla_l u_k u'_j-c_{ijkl}\nabla_l u'_k u_j]dS_i\end{aligned} \tag{2.250}$$

where we assume that the integrals converge. In equation (2.250) $d\mathbf{S}$ is a vector pointing out of the volume V. If the volume V is unbounded, a radiation condition must be invoked (see Tan 1975). Now if the equations of motion are satisfied

$$\nabla_i(c_{ijkl}\nabla_l u_k) = \rho(u_{j,tt}-f_j) \tag{2.251}$$

and similarly for $\mathbf{u}'$. Then

$$\begin{aligned}\int_V \rho\left[(u_{j,tt}-f_j)u'_j-(u'_{j,tt}-f'_j)u_j\right]dV \\ = \int_S \left[c_{ijkl}\nabla_l u_k u'_j - c_{ijkl}\nabla_l u'_k u_j\right]dS_i\cdot\end{aligned} \tag{2.252}$$

The correspondence of equation (2.252) with the statement of the reciprocity principle is made clearer if we rewrite it in vector notation, and substitute strains for the derivatives of the displacements. Making use of the symmetry properties of the elastic constants, we obtain

$$\int_V \rho\left[(\mathbf{u}_{,tt}-\mathbf{f})\cdot\mathbf{u}'-(\mathbf{u}'_{,tt}-\mathbf{f}')\cdot\mathbf{u}\right]dV = \int_S \left[\mathbf{u}'\cdot\boldsymbol{\tau}-\mathbf{u}\cdot\boldsymbol{\tau}'\right]\cdot d\mathbf{S}. \tag{2.253}$$

Now the terms on the left-hand side of equation (2.253) represent the difference between the work done in increasing the kinetic energy and the work done by the body forces. The terms on the right-hand side represent the work done by tractions on the surface.

We can now exploit the generality of (2.253) by specifying that the force $\mathbf{f}'$ shall act at a point and instantaneously, in other words, it is an instantaneous delta function point force:

$$\rho f'_i = \delta_{ij}\delta(t)\delta(\mathbf{r}-\mathbf{r}'). \tag{2.254}$$

The response to this force is the Green function for the problem.

Further development is simplified if we assume a single-frequency time variation, when the time derivatives in equation (2.253) give equal and opposite terms, and we obtain an integral involving the single-frequency Green tensor $G_{ij}(\mathbf{r}|\mathbf{r}')$. If both the stress and the displacement are regular within a volume V bounded by a surface S then

$$\begin{aligned}&\int_S \left\{\tau_{ij}(\mathbf{r}')G_{ik}(\mathbf{r}|\mathbf{r}')\hat{n}_j - u_i(\mathbf{r}')\hat{n}_j\Sigma_{ijk}(\mathbf{r}|\mathbf{r}')\right\}dS' \\ &+ \int_V \rho f_j(\mathbf{r}')G_{jk}(\mathbf{r}|\mathbf{r}')dV' = \begin{cases} u_k(\mathbf{r}) & \mathbf{r}\text{ inside } V \\ 0 & \mathbf{r}\text{ outside } V\end{cases}\end{aligned} \tag{2.255}$$

where $\hat{\mathbf{n}}$ is the unit vector normal to the surface S and pointing from S out of the volume V, that is, it is the vectorial part of the surface element $d\mathbf{S}$, and $\Sigma_{ijk}(\mathbf{r}|\mathbf{r}')$ is the stress Green tensor. The stress Green tensor is related to G_{ik} in the usual way

$$\Sigma_{ijk}(\mathbf{r}|\mathbf{r}') = c_{ijkl}\frac{\partial}{\partial r_m}G_{ml}(\mathbf{r}|\mathbf{r}'), \tag{2.256}$$

or, for isotropic materials,

$$\Sigma_{ijk} = \lambda\delta_{ij}\frac{\partial}{\partial r_m}G_{mk}(\mathbf{r}|\mathbf{r}')+\mu\left[\frac{\partial}{\partial r_i}G_{jk}(\mathbf{r}|\mathbf{r}') + \frac{\partial}{\partial r_j}G_{ik}(\mathbf{r}|\mathbf{r}')\right]. \tag{2.257}$$

Equation (2.255) is the important result of this section, which represents, in principle, the solution of any elastodynamic problem.

The Green tensor on which it depends, $G_{ik}(\mathbf{r}|\mathbf{r}')$, represents the displacement in the direction k at $\mathbf{r}$ caused by a unit point force at $\mathbf{r}'$ directed in the direction i.

The Green tensor may, as shown by Morse and Feshbach, be built up from the Green function for scalar waves $g(\mathbf{r}|\mathbf{r}')$. In three dimensions this scalar form is

$$g(\mathbf{r}|\mathbf{r}') = \frac{e^{i\omega|\mathbf{r}-\mathbf{r}'|/c}}{|\mathbf{r}-\mathbf{r}'|}. \tag{2.258}$$

Alternatively, one may work directly from the elastodynamic equations. Tan (1975) gave a straightforward derivation of the form of the Green function, based on a transformation from $(\mathbf{r},t)$ space to $(\mathbf{k},\omega)$ space using Fourier transforms. The transform $\tilde{f}$ of a function f is

$$\tilde{f}(\mathbf{k},\omega) \equiv \int\int\int\int f(\mathbf{r},t)e^{i(\omega t-\mathbf{k}\cdot\mathbf{r})}\,d\mathbf{r}dt, \tag{2.259}$$

with the corresponding inversion expression

$$\frac{1}{(2\pi)^4}f(\mathbf{r},t) \equiv \int\int\int\int \tilde{f}(\mathbf{k},\omega)e^{-i(\omega t-\mathbf{k}\cdot\mathbf{r})}\,d\mathbf{k}d\omega. \tag{2.260}$$

We assume that the function f is such that the conditions of the Fourier inversion theorem are met: further details may be found in standard texts such as Titchmarsh (1948).

Substituting in the equation satisfied by the Green tensor, namely the elastodynamic equation with a delta function force, we obtain

$$c_{ijkl}k_jk_l\tilde{G}_{km}(\mathbf{k}|\mathbf{r}')-\rho\omega^2\tilde{G}_{im}(\mathbf{k}|\mathbf{r}') = \delta_{im}e^{-i\mathbf{k}\cdot\mathbf{r}'} \tag{2.261}$$

which is a matrix equation for $\tilde{\mathbf{G}}$. The presence of the exponential on the right-hand side ensures that when the inversion theorem is applied $\mathbf{G}$ will depend only on the difference $\mathbf{r}-\mathbf{r}'$. Solution of the matrix equation is complicated except for an isotropic medium, when

$$\tilde{G}_{ij}(\mathbf{k}|\mathbf{r}') = -\frac{(\lambda+2\mu)(k_p^2-k^2)\delta_{ij}+(\lambda+\mu)k_ik_j}{\mu(\lambda+2\mu)(k_p^2-k^2)(k_s^2-k^2)}e^{-i\mathbf{k}\cdot\mathbf{r}'}. \tag{2.262}$$

Here we have introduced the usual compression and shear wavenumbers. Equation (2.262) may be rewritten as

$$\tilde{G}_{ij}(\mathbf{k}|\mathbf{r}') = -\frac{\delta_{ij}}{\mu(k_p^2-k^2)}e^{-i\mathbf{k}\cdot\mathbf{r}'} - \frac{1}{\rho\omega^2}\left[\frac{k_ik_j}{k_p^2-k^2} - \frac{k_ik_j}{(k_s^2-k^2)}\right]e^{-i\mathbf{k}\cdot\mathbf{r}'}. \tag{2.263}$$

The basic integral in the Fourier inversion now has the form

$$I(\mathbf{R}) = \frac{1}{(2\pi)^4}\int\frac{e^{i\mathbf{k}\cdot\mathbf{R}}}{q^2-k^2}d\mathbf{k}\cdot \tag{2.264}$$

In order to evaluate this integral we have to assume that q has a small imaginary part, so as to move the poles of the integrand away from the path of integration along the real axis. In physical terms this is usually justified on the grounds that no material is totally lossless, and the imaginary parts represent dissipation in the medium. In fact, only a positive imaginary part represents dissipation: a negative imaginary part would lead to exponential growth of an outgoing spherical wave. In the limit as $I\ (q) \rightarrow 0^+$, we obtain

$$I(\mathbf{R}) = \frac{-1}{4\pi} \frac{e^{ik_p R}}{R}. \tag{2.265}$$

This result may be used to evaluate the Fourier inversion of equation (2.263), and the resulting expression for the elastodynamic Green tensor is

$$G_{ik}(\mathbf{r}|\mathbf{r}') = \frac{1}{4\pi\rho\omega^2}\left\{\delta_{ik}k_s^2\frac{e^{ik_s R}}{R} - \frac{\partial}{\partial r_i}\frac{\partial}{\partial_k}\left[\frac{e^{ik_p R}}{R} - \frac{e^{ik_s R}}{R}\right]\right\}, \tag{2.266}$$

where now the relative distance $R = |\mathbf{r}-\mathbf{r}'|$.

To obtain the Green function for transient waves, one takes the Fourier transform of the continuous-wave equations. As the frequency-space Green function depends on frequency, the function in the time domain depends on time differences: we therefore introduce $T \equiv t-t'$. Then the integral equations become

$$\begin{aligned}
&\int_0^{t^+} dt' \int_S dS' \{\hat{n}_j \tau_{ij}(\mathbf{r}',t')G_{ik}(\mathbf{R},T) - u_i(\mathbf{r}',t')\hat{n}_j\Sigma_{ijk}(\mathbf{R},T)\} + \\
&+\int_0^{t^+} dt' \int_V dV' f_j(\mathbf{r}',t')G_{jk}(\mathbf{R},T)dV' \\
&+\int_V dV'\rho\left[G_{jk}(\mathbf{R},T)\frac{\partial}{\partial t'}u_j(\mathbf{r}',t') - u_j(\mathbf{r}',t')\frac{\partial}{\partial t'}G_{jk}(\mathbf{R},T)\right]_{t'=0} \\
&\qquad = \begin{cases} u_k(\mathbf{r},t) & \mathbf{r} \text{ inside } V \\ 0 & \mathbf{r} \text{ outside } V. \end{cases}
\end{aligned} \tag{2.267}$$

We have introduced the notation t^+ is used to denote a time infinitesimally greater than t; it is required to avoid ending an integration on the peak of a delta function. The time-domain Green function is, in three dimensions,

$$\begin{aligned}
G_{ij}(\mathbf{R},T) = &\frac{1}{4\pi\rho R}\left\{\hat{R}_i\hat{R}_j\frac{\delta(T-R/c_p)}{c_p^2} + (\delta_{ij}-\hat{R}_i\hat{R}_j)\frac{\delta(T-R/c_s)}{c_s^2}\right\} \\
&-\frac{1}{4\pi\rho R^3}(\delta_{ij}-3\hat{R}_i\hat{R}_j)\int_{R/c_p}^{R/c_s} T'\delta(T-T')dT'.
\end{aligned} \tag{2.268}$$

Here the separate compression and shear wave contributions are clear, but there is also a term, in the integral, which is dependent on the initial values of the displacements. The disturbance which arises from this integral arrives at a time between the compression and shear wave arrival times. A general point source, then, if excited with a delta function time variation, will give rise to displacements a time T later which are confined to a spherical shell bounded by spheres of radii $c_s T$ and $c_p T$.

Equation (2.268) may be interpreted in terms of Huygens' principle. The two terms involving delta functions show that any initial dilatation will give rise to a wavefront travelling with the compression wave speed, and initial shear will give rise to a shear wave. The integral, however, does not have an analogy in the more familiar scalar wave form of Huygens' principle.

The Green tensor is not in itself a panacea, as equation (2.255) is still an integral equation for the field **u**, and there is no reason to expect it to be easier to solve than the differential form of the wave equation. The integral form of the equation is, however, an important starting point for systematic approximations to the solutions of scattering problems, as described in the next chapter. It is quite straightforward to derive asymptotic fields using the Green function, and this allows total scattering cross-sections to be investigated.

CHAPTER 3
Methods of solution

For nondestructive testing applications, we are interested in the entire process of elastic wave generation, propagation, scattering and detection. If we are concerned with linear elasticity, it is possible to treat the scattering problem in isolation. It is therefore usual to treat the scattering of infinite plane waves, and to put aside such effects as the shape of the beams from the transducers. We follow this approach in the analytic treatments of this chapter, though it should be noted that purely numerical methods will generally work with finite beams. If the results of the analytic approaches are to be taken as models of actual scattering events in nondestructive testing it should be recognized that they apply only when the scatterer is illuminated by a plane wave and is in the far field of an unfocused detecting transducer: for other cases linear superpositions of solutions must be taken. This point has been discussed by Kino (1978), and he shows that the amplitude of the signal detected when ultrasound is scattered from a small flaw can vary as the frequency, its square or its cube depending on the distances of the flaw from the generating and detecting transducers. Kino also suggests how reference scatterers may be used for normalizing purposes, to eliminate the characteristics of the transducers and isolate the effects of the scatterering itself. Reference scatterers are used in a similar way by other groups (for example, Coffey and Chapman 1983). We shall return to the problem of including transducer characteristics in Chapter 5: in the present chapter we consider plane wave excitations.

3.1 Exact solutions

It is an unfortunate fact that there are very few problems in scalar wave theory which are exactly soluble, and solutions of the elastodynamic equations are slightly harder. There are two difficulties which may be described in terms of a solution in the form of an eigenfunction expansion. The scalar wave equation in three dimensions depends on three coordinates, but we are free to transform the coordinate system. In general, we take the three curvilinear coordinates $\boldsymbol{\xi}$ with a set of scale factors $\mathbf{h}$, so that the length scale is defined by

$$ds^2 = dr_i dr_i = h_{ij} \xi_j h_{ik} \xi_k . \tag{3.1}$$

In an orthogonal curvilinear system, $\mathbf{h}$ does not include any cross terms, so

$$h_{ij} = h_i \delta_{ij}, \tag{3.2}$$

and the scale factors are computed from the coordinate transformation by

$$h_i^2 = \sum_{j=1}^{3} \left(\frac{\partial r_j}{\partial \xi_i}\right)^2 \left[\sum_{j=1}^{3} \left(\frac{\partial \xi_j}{\partial r_i}\right)^2\right]^{-1} . \tag{3.3}$$

Now there are only eleven coordinate systems for which the scalar wave equation is separable (Morse and Feshbach 1953, Chapter 5). By separable is meant that if the solution is written as a product

$$f(\xi) = \Xi_1(\xi_1)\Xi_2(\xi_2)\Xi_3(\xi_3) \tag{3.4}$$

then the differential equation separates into three terms, each of which is a function of only one of the components ξ_i. Specifically, the systems are the general ellipsoidal coordinate system or the various degenerate forms of it: rectangular; circular cylindrical; elliptic cylindrical; parabolic cylindrical; spherical; conical; parabolic; prolate spheroidal; oblate spheroidal; ellipsoidal; paraboloidal. Solutions to the scalar wave equation are possible when the surfaces on which the boundary conditions are specified correspond to constant-coordinate surfaces.

Orthogonal curvilinear coordinates

System	Coordinates	Special functions
Rectangular	r_1 (x) r_2 (y) r_3 (z)	$\cos/\sin(k_i r_i)$ or $e^{ik_i r_i}$
Circular cylindrical	r θ z $\theta = \tan^{-1}(y/x)$	Bessel function $\cos/\sin(n\theta)$ $\cos/\sin(k_z z)$
Elliptic cylindrical	$\xi_1 + i\xi_2 = \cosh^{-1}((x+iy)/a)$ z $2a$ is the focal length of the ellipse.	Modified Mathieu function Periodic Mathieu function $\cos/\sin(k_z z)$
Parabolic cylindrical	$\xi_1 + i\xi_2 = \sqrt{(2(x+iy))}$ z	Weber function $\cos/\sin(k_z z)$
Spherical	r θ ϕ $\phi = \tan^{-1}(y/x)$ $\theta = \tan^{-1}(\sqrt{(x^2+y^2)}/x)$	Spherical Bessel function Legendre polynomial $\cos/\sin(n\phi)$

One way of proceeding is to expand each of the functions Ξ_i in terms of the eigenfunctions Λ_n of the three ordinary differentials. These eigenfunctions are generally special functions whose properties are well known and may be found in reference works (for example, Abramowitz and Stegun 1965). The boundary conditions then reduce to simple relationships between the coefficients in the expansions, and the problem may be solved to any desired accuracy by continuing the expansion to sufficiently high order n. This procedure will be illustrated below.

Figure 20

The constant coordinate surfaces for the circular cylindrical coordinate system.

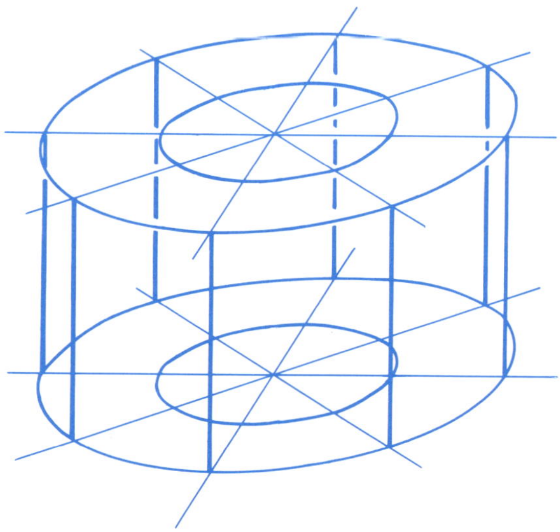

For the vector wave equation, a more restricted set of coordinate systems allows exact solutions. The problem concerns the resolution into potentials, where the vector potentials will only be determined by wave equations if at least one of the scale factors h_i is unity and the ratio of the other two scale factors is independent of the coordinate ξ_i. The wavefronts of transverse waves are then chosen to correspond to surfaces of constant ξ_i and the separation is possible. Only the first six of the systems listed above have this property: the systems, with the appropriate transformations and the preferred axes are given in the table. The table also lists the special functions which are involved in solutions in each system. The form of the constant coordinate surfaces is shown in Figures 20 to 23. Details of the conical coordinate system, for which the transformations involve elliptic functions, are omitted. Problems in parabolic, oblate spheroidal and prolate spheroidal

Figure 21

The constant coordinate surfaces for the elliptic cylindrical system.

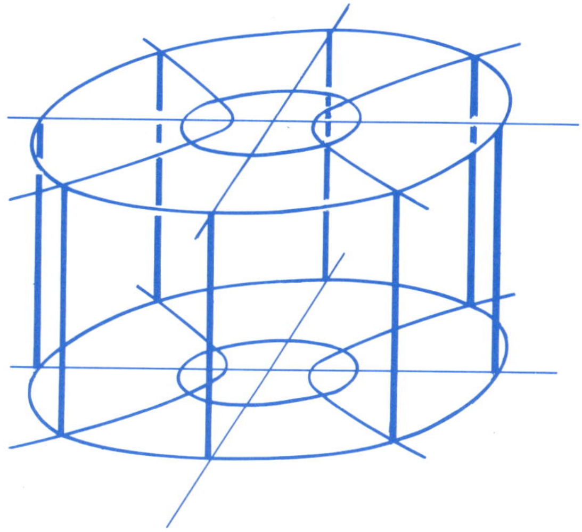

Figure 22

The constant coordinate surfaces for the circular parabolic cylindrical system.

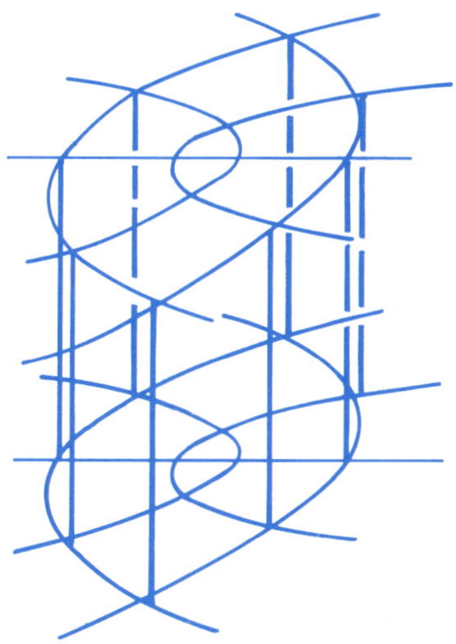

Figure 23

The constant coordinate surfaces for the spherical system.

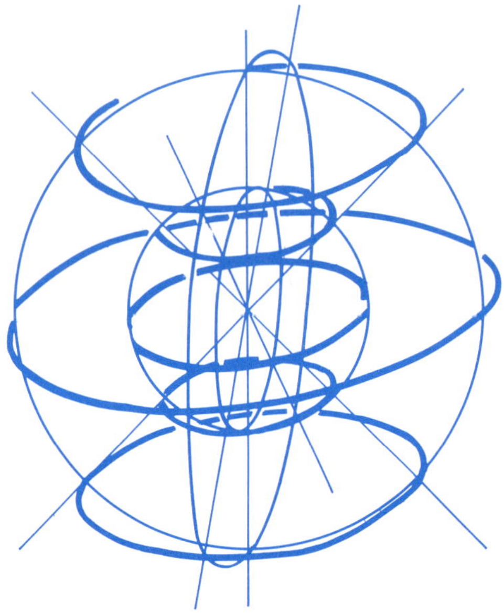

coordinates simplify if they have cylindrical symmetry, and then become separable because only a single vector potential is required.

For the elastodynamic equation, there is a further complication. The boundary conditions, as we have seen, can cause mixing of compression and shear waves. Similarly, they cause mixing between eigenfunctions in all the coordinate systems except rectangular, circular cylindrical and spherical. The coefficients for different orders n are then coupled together, and an infinite set of simultaneous equations must be solved to find the solution.

Figures 20 to 23 give an idea of the types of problem that can be solved exactly. Of particular interest are the degenerate cases of the coordinate systems. For the elliptic cylinder the limiting case is a line joining the foci of an ellipse, and imposing boundary conditions on such a surface allows one to treat the scattering from a ribbon-like crack. The degenerate parabola is a line from the origin to infinity, which may be taken to represent a semi-infinite crack. Exact solutions are possible, then, for these two cases of interest in nondestructive testing. It is often difficult to use such solutions, however, because tabulations of the special functions required are sparse, and few standard subroutine libraries for computers contain routines for Mathieu functions and Weber functions.

We shall treat here two cases which can be solved using the expansion method, one in spherical coordinates and one in cylindrical coordinates. The special functions required for these cases are readily available. The cases we select concern the scattering of elastic waves by a spherical inclusion and the scattering of elastic waves by a cylindrical inclusion, These cases are intended to fulfil several functions: they illustrate the expansion method, they permit comparison with other analytic techniques, and they provide precise theories for comparison with numerical techniques in the following chapter. Further details of analytical methods and eigenfunction expansions in particular are given by Pao and Mow (1973).

3.1.1 Elastic waves in spherical coordinates

The scattering of elastic waves by a spherical inclusion is a model for several problems of practical interest. As an example, suppose that a metallic structure has been welded: in the course of welding it is possible that slag and gas bubbles have been incorporated in the material. A nondestructive testing technique might be required to distinguish between these two types of inclusion. A full history of this scattering problem is given by Pao and Mow (1973), but recent work dates back to work by Ying and Truell (1956), Einspruch and Truell (1960), Einspruch, Witterholt and Truell (1960), and Knopoff (1959a, b). The literature, unfortunately, contains several misprints, some of which are corrected by Gubernatis *et al.* (1977).

The transformation to spherical polar coordinates is defined by taking

$$\xi_1 = r \qquad \xi_2 = \theta \qquad \xi_3 = \phi \tag{3.5}$$

where

$$x = r\cos\phi\sin\theta \quad y = r\sin\phi\sin\theta \quad z = r\cos\theta \tag{3.6}$$

and, using the subscripts r,θ,ϕ instead of 1,2,3 for clarity, the scale factors are

$$h_r = 1 \quad h_\theta = r \quad h_\phi = r\sin\theta. \tag{3.7}$$

We shall require the following standard forms for spherical polar coordinates:

$$\nabla f = \hat{\mathbf{r}}\frac{\partial f}{\partial r} + \hat{\boldsymbol{\theta}}\frac{\partial f}{\partial \theta} + \hat{\boldsymbol{\phi}}\frac{1}{\sin\theta}\frac{\partial f}{\partial \phi}; \tag{3.8}$$

$$\nabla\cdot\mathbf{f} = \frac{1}{r^2}\frac{\partial}{\partial r}(r^2 f_r) + \frac{1}{r\sin\theta}\frac{\partial}{\partial\theta}(\sin\theta\, f_\theta) + \frac{1}{r\sin\theta}\frac{\partial f}{\partial\phi_\phi}; \tag{3.9}$$

$$\nabla_\wedge \mathbf{f} = \hat{\mathbf{r}}\frac{1}{r\sin\theta}\left[\frac{\partial}{\partial\theta}(\sin\theta\, f_\phi) - \frac{\partial}{\partial\phi}f_\theta\right] + \hat{\boldsymbol{\theta}}\frac{1}{r}\left[\frac{1}{\sin\theta}\frac{\partial}{\partial\phi}f_r - \frac{\partial}{\partial r}(rf_\phi)\right] + \hat{\boldsymbol{\phi}}\frac{1}{r}\left[\frac{\partial}{\partial r}(rf_\theta) - \frac{\partial}{\partial\theta}f_r\right]; \tag{3.10}$$

$$\nabla^2 f = \frac{1}{r^2}\frac{\partial}{\partial r}\left(r^2\frac{\partial f}{\partial r}\right) + \frac{1}{r^2\sin\theta}\frac{\partial}{\partial\theta}\left(\sin\theta\frac{\partial f}{\partial\theta}\right) + \frac{1}{r^2\sin^2\theta}\frac{\partial^2 f}{\partial\phi^2}. \tag{3.11}$$

Three potentials ϕ, ψ and χ may be defined, each of which satisfies the wave equation in the form

$$\frac{1}{r^2}\frac{\partial}{\partial r}\left(r^2\frac{\partial f}{\partial r}\right) + \frac{1}{r^2\sin\theta}\frac{\partial}{\partial\theta}\left(\sin\theta\frac{\partial f}{\partial\theta}\right) + \frac{1}{r^2\sin\theta}\frac{\partial^2 f}{\partial\phi^2} = \frac{1}{c^2}\frac{\partial^2 f}{\partial t^2}, \tag{3.12}$$

with $c = c_p$ for ϕ, $c = c_s$ for ψ and χ, and the displacements given by

$$u_r = \frac{\partial\phi}{\partial r} + \frac{\partial^2(r\chi)}{\partial r^2} - r\nabla^2\chi \tag{3.13}$$

$$u_\theta = \frac{1}{r}\frac{\partial\phi}{\partial\theta} + \frac{1}{r\sin\theta}\frac{\partial}{\partial\phi}(r\psi) + \frac{1}{r}\frac{\partial^2(r\chi)}{\partial\theta\partial r} \tag{3.14}$$

$$u_\phi = \frac{1}{r\sin\theta}\frac{\partial\phi}{\partial\phi} - \frac{1}{r}\frac{\partial}{\partial\theta}(r\psi) + \frac{1}{r\sin\theta}\frac{\partial^2(r\chi)}{\partial\phi\partial r}. \tag{3.15}$$

It should be noted that the potential χ has different dimensions from the other two, containing an extra factor of a length. The relationships between strain and displacement in spherical polar coordinates are

$$e_{rr} = \frac{\partial}{\partial r}u_r \tag{3.16}$$

$$e_{\theta\theta} = \frac{1}{r}\frac{\partial}{\partial\theta}u_\theta + \frac{1}{r}u_r \tag{3.17}$$

$$e_{\phi\phi} = \frac{1}{r\sin\theta}\frac{\partial}{\partial\phi}u_\phi + \frac{1}{r}u_r + \frac{1}{r}\cot\theta\, u_\theta \tag{3.18}$$

$$e_{r\theta} = \frac{1}{2r}\frac{\partial}{\partial\theta}u_r - \frac{1}{2r}u_\theta + \frac{1}{2}\frac{\partial}{\partial r}u_\theta \tag{3.19}$$

$$e_{r\phi} = \frac{1}{2r\sin\theta}\frac{\partial}{\partial\phi}u_r - \frac{1}{2r}u_\phi + \frac{1}{2}\frac{\partial}{\partial r}u_\phi \tag{3.20}$$

$$e_{\theta\phi} = \frac{1}{2r}\frac{\partial}{\partial\theta}u_\phi - \frac{\cot\theta}{2r}u_\phi + \frac{1}{2r\sin\theta}\frac{\partial}{\partial\phi}u_\theta. \tag{3.21}$$

When the elastic constants are included, to convert the strains to stresses, and the potentials substituted into the above strain formulae, the expressions become quite lengthy:

$$\tau_{rr} = \lambda\nabla^2\phi + 2\mu\frac{\partial^2\phi}{\partial r^2} + 2\mu\frac{\partial}{\partial r}\left[\frac{\partial^2(r\chi)}{\partial r^2} - r\nabla^2\chi\right] \tag{3.22}$$

$$\begin{aligned}\tau_{\theta\theta} = {} & \lambda\nabla^2\phi + 2\mu\left[\frac{1}{r}\frac{\partial\phi}{\partial r} + \frac{1}{r^2}\frac{\partial^2\phi}{\partial\theta^2}\right] + 2\mu\frac{1}{r}\frac{\partial}{\partial\theta}\left[\frac{1}{r\sin\theta}\frac{\partial(r\psi)}{\partial\phi}\right] \\ & + 2\mu\left[\frac{1}{r^2}\frac{\partial^3(r\chi)}{\partial\theta^2\partial r} + \frac{1}{r}\frac{\partial^2(r\chi)}{\partial r^2} - \nabla^2\chi\right]\end{aligned} \tag{3.23}$$

$$\begin{aligned}\tau_{\phi\phi} = {} & \lambda\nabla^2\phi + 2\mu\left[\frac{1}{r\sin\theta}\frac{\partial}{\partial\phi}\left(\frac{1}{r\sin\theta}\frac{\partial\phi}{\partial\phi}\right) + \frac{1}{r}\frac{\partial\phi}{\partial r} + \frac{1}{r^2}\cot\theta\frac{\partial\phi}{\partial\theta}\right] \\ & + 2\mu\left[\frac{\cos\theta}{r^2\sin^2\theta}\frac{\partial(r\psi)}{\partial\phi} - \frac{1}{r^2\sin\theta}\frac{\partial^2(r\psi)}{\partial\phi\partial\theta}\right] \\ & + 2\mu\left[\frac{1}{r^2\sin^2\theta}\frac{\partial^3(r\chi)}{\partial\phi^2\partial r} + \frac{1}{r}\frac{\partial^2(r\chi)}{\partial r^2} - \nabla^2\chi + \frac{\cos\theta}{r^2\sin\theta}\frac{\partial^2(r\chi)}{\partial\theta\partial r}\right]\end{aligned} \tag{3.24}$$

$$\begin{aligned}\tau_{r\theta} = {} & 2\mu\left[\frac{1}{r}\frac{\partial^2\phi}{\partial r\partial\theta} - \frac{1}{r^2}\frac{\partial\phi}{\partial\theta}\right] - \mu\left[\frac{1}{r^2\sin\theta}\frac{\partial(r\psi)}{\partial\phi} - \frac{\partial}{\partial r}\left(\frac{1}{r\sin\theta}\frac{\partial(r\psi)}{\partial\phi}\right)\right] \\ & + \mu\left[\frac{1}{r}\frac{\partial}{\partial\theta}\left(\frac{\partial^2(r\chi)}{\partial r^2} - r\nabla^2\chi\right) - \frac{1}{r^2}\frac{\partial^2(r\chi)}{\partial r\partial\theta} + \frac{\partial}{\partial r}\left(\frac{1}{r}\frac{\partial^2(r\chi)}{\partial r\partial\theta}\right)\right]\end{aligned} \tag{3.25}$$

$$\begin{aligned}\tau_{r\phi} = {} & 2\mu\left[\frac{1}{r\sin\theta}\frac{\partial^2\phi}{\partial r\partial\phi} - \frac{1}{r^2\sin\theta}\frac{\partial\phi}{\partial\phi}\right] \\ & + \mu\left[\frac{2}{r^2}\frac{\partial(r\psi)}{\partial\theta} - \frac{1}{r}\frac{\partial^2(r\psi)}{\partial r\partial\theta}\right] \\ & + \mu\left[\frac{1}{r\sin\theta}\frac{\partial}{\partial\phi}\left(2\frac{\partial^2(r\chi)}{\partial r^2} - r\nabla^2\chi\right) - \frac{2}{r^2\sin\theta}\frac{\partial^2(r\chi)}{\partial r\partial\phi}\right]\end{aligned} \tag{3.26}$$

$$\begin{aligned}\tau_{\theta\phi} = {} & 2\mu\left[\frac{1}{r^2\sin\theta}\frac{\partial^2\phi}{\partial\theta\partial\phi} - \frac{\cos\theta}{r^2\sin^2\theta}\frac{\partial\phi}{\partial\phi}\right] \\ & + \mu\left[\frac{\cos\theta}{r^2\sin\theta}\frac{\partial(r\psi)}{\partial\theta} - \frac{1}{r^2}\frac{\partial^2(r\psi)}{\partial\theta^2} + \frac{1}{r^2\sin^2\theta}\frac{\partial^2(r\psi)}{\partial\phi^2}\right] \\ & + 2\mu\left[\frac{1}{r^2\sin\theta}\frac{\partial^3(r\chi)}{\partial r\partial\theta\partial\phi} - \frac{\cos\theta}{r^2\sin^2\theta}\frac{\partial^2(r\chi)}{\partial r\partial\phi}\right].\end{aligned} \tag{3.27}$$

Before proceeding further, we should examine the wave equation in spherical coordinates and find the eigenfunctions which are appropriate for expansions. If we consider a single frequency, the time derivative reduces to a multiplication by $-\omega^2$, and the equation reduces to the

Helmholtz equation:

$$\frac{1}{r^2}\frac{\partial}{\partial r}(r^2\frac{\partial f}{\partial r}) + \frac{1}{r^2\sin\theta}\frac{\partial}{\partial\theta}(\sin\theta\frac{\partial f}{\partial\theta}) + \frac{1}{r^2\sin\theta}\frac{\partial^2 f}{\partial\phi^2} + \frac{\omega^2}{c^2}f = 0, \quad (3.28)$$

which may be separated by writing

$$f(r,\theta,\phi) = R(r)\Theta(\theta)\Phi(\phi). \quad (3.29)$$

The separated equations then become

$$r^2\frac{d^2R}{dr^2} + 2r\frac{dR}{dr} + (k^2r^2 - l(l+1))R = 0 \quad (3.30)$$

$$\frac{1}{\sin\theta}\frac{d}{d\theta}\left(\sin\theta\frac{d\Theta}{d\theta}\right) + \left((l(l+1) - \frac{m^2}{\sin^2\theta}\right)\Theta = 0 \quad (3.31)$$

$$\frac{d^2\Phi}{d\phi^2} + m^2\Phi = 0 \quad (3.32)$$

where the separation constants $l(l+1)$ and m^2 have been given forms which simplify the following expressions, and k is as usual equal to ω/c.

The azimuthal variation, from equation (3.32), has the form of a complex exponential

$$\Phi(\phi) = e^{\pm im\phi} \quad (3.33)$$

and m must be an integer if the solution is to be single-valued.

The equation in θ may be rewritten in the form of the associated Legendre equation by defining $c \equiv \cos\theta$, when

$$(1 - c^2)\Theta'' - 2c\Theta' + \left[l(l+1) - \frac{m^2}{(1 - c^2)}\right]\Theta = 0, \quad (3.34)$$

in which primes denote differentiation with respect to c. Of the two sets of solutions to this equation, P_{lm} and Q_{lm}, Q_{lm} is singular at $c = \pm 1$, and so we reject this family. P_{lm} converges at $c = \pm 1$ if l is an integer, and then P_{lm} is a polynomial in c, known as the associated Legendre polynomial. Its properties are well documented in standard reference works. The product of Φ and Θ, then, is the familiar spherical harmonic

$$Y_{lm}(\theta,\phi) = \mathrm{P}_{lm}(cos\theta)e^{im\phi}, \quad (3.35)$$

although Y_{lm} is usually written with a normalizing factor which ensures

$$\textstyle\int_0^{2\pi}\int_0^{\pi} Y^*_{lm}(\theta,\phi)Y_{LM}(\theta,\phi)d\theta d\phi = \delta_{lL}\delta_{mM}. \quad (3.36)$$

The recurrence relations linking different l and m values of P_{lm} are as follows:

$$(1-c^{2)}\frac{dP_{lm}(c)}{dc} = (l+m)P_{l-1\,m}(c) - lP_{lm}(c) \tag{3.37}$$

$$(1-c^{2)}\frac{dP_{lm}(c)}{dc} = (l+1)cP_{lm}(c) - (l-m+1)P_{l+1\,m}(c) \tag{3.38}$$

$$(2l+1)cP_{lm}(\cos\theta) = (l-m+1)P_{l+1\,m}(c) + (l+m)P_{l-1\,m}(c). \tag{3.39}$$

The radial equation (3.30) can be transformed into a Bessel equation of half-integer order. We define spherical Bessel functions j_n, y_n, $h_n^{(1)}$ and $h_n^{(2)}$ by analogy with the cylindrical Bessel functions J_n, Y_n, $H_n^{(1)}$ and $H_n^{(2)}$ by using

$$\mathfrak{S}_n(kr) \equiv \sqrt{\frac{\pi}{2kr}}\,\mathfrak{C}_{n+\frac{1}{2}}(kr) \tag{3.40}$$

in which $\mathfrak{S}$ denotes one of the spherical functions and $\mathfrak{C}$ one of the cylindrical functions. $R(r)$ transformed in this way,

$$S(r) \equiv \frac{1}{\sqrt{kr}}R(r), \tag{3.41}$$

satisfies the Bessel equation

$$r^2\frac{d^2S}{dr^2} + 2r\frac{dS}{dr} + (k^2r^2 - (l+\tfrac{1}{2})^2)S = 0, \tag{3.42}$$

and the solutions of (3.42) are the Bessel functions of order $n+\frac{1}{2}$.

In contrast with the cylindrical functions, the spherical Bessel functions can be expressed as simple combinations of algebraic and trigonometric functions. Useful expressions are

$$j_n(z) = (-1)^n z^z\left(\frac{1}{z}\frac{d}{dz}\right)^n\frac{\sin z}{z} \tag{3.43}$$

$$y_n(z) = -(-1)^n z^z\left(\frac{1}{z}\frac{d}{dz}\right)^n\frac{\cos z}{z} \tag{3.44}$$

$$h_n^{(1)}(z) = -i(-1)^n z^z\left(\frac{1}{z}\frac{d}{dz}\right)^n\frac{e^{iz}}{z} \tag{3.45}$$

$$h_n^{(2)}(z) = i(-1)^n z^z\left(\frac{1}{z}\frac{d}{dz}\right)^n\frac{e^{-iz}}{z}, \tag{3.46}$$

so that the low order functions have particularly simple forms:

$$j_0(z) = \frac{\sin z}{z} \tag{3.47}$$

$$y_0(z) = \frac{\cos z}{z} \tag{3.48}$$

$$h_0^{(1)}(z) = -i\frac{e^{iz}}{z} \tag{3.49}$$

$$h_0^{(2)}(z) = i\frac{e^{-iz}}{z} \tag{3.50}$$

$$j_1(z) = (\sin z - z\cos z)/z^2 \tag{3.51}$$

$$y_1(z) = -(\cos z + z\sin z)/z^2 \tag{3.52}$$

$$h_1^{(1)}(z) = i(iz - 1)e^{iz}/z^2 \tag{3.53}$$

$$h_1^{(2)}(z) = -i(-iz - 1)e^{-iz}/z^2. \tag{3.54}$$

The other properties of the Bessel functions which we require are the recurrence relations, which relate functions of different order to one another:

$$\frac{2n+1}{z}\mathfrak{S}_n(z) = \mathfrak{S}_{n-1}(z) + \mathfrak{S}_{n+1}(z) \tag{3.55}$$

$$(2n+1)\frac{d}{dz}\mathfrak{S}_n(z) = n\mathfrak{S}_{n-1}(z) - (n+1)\mathfrak{S}_{n+1}(z) \tag{3.56}$$

$$\frac{d}{dz}\left(z^{n+1}\mathfrak{S}_n(z)\right) = z^{n+1}\mathfrak{S}_{n-1}(z) \tag{3.57}$$

$$\frac{d}{dz}\left(z^{-n}\mathfrak{S}_n(z)\right) = -z^{-n}\mathfrak{S}_{n+1}(z). \tag{3.58}$$

Before proceeding further, we should note some of the properties of the spherical Bessel functions which will be important in selecting them for expanding the solutions of wave scattering problems. For large arguments, the complex functions $h_n^{(1),(2)}(kr)$ tend to incoming or outgoing radial waves

$$h_n^{(1),(2)} \sim \frac{e^{\pm ikr}}{kr} \tag{3.59}$$

and they are therefore suitable for expansions in the region remote from a scatterer. The real functions, on the other hand, resemble standing waves, and might both be candidates for the expansion of the field inside a spherical impurity. Only j_n is bounded at $r=0$, however, so expansions inside the scatterer are made in terms of j_n alone. For small arguments, the spherical Bessel functions vary steadily with increasing order:

$$j_n(z) \rightarrow \frac{z^n}{1.3.5\ldots(2n+1)} \tag{3.60}$$

$$y_n(z) \rightarrow \frac{z^{-n-1}}{1.3.5\ldots(2n-1)}, \tag{3.61}$$

which will be of use in analysing the scattering from small spherical inclusions.

3.1.2 Scattering of compression waves by a sphere

By choosing a plane compression wave as the incident wave, the problem simplifies because it acquires axial symmetry about the incoming wavevector. The dependence on the azimuthal angle ϕ vanishes, and the angular part of the fields is determined by P_l rather than by Y_{lm}. We can then write down expansions of the wavefield, bearing in mind the considerations of the end of the previous section.

The spherical inclusion may have different elastic constants or density from the surrounding medium, and its properties are distinguished by a prime notation. The time variation as $\exp(-i\omega t)$ is omitted throughout. Outside the sphere the scattered wave must be outgoing,

$$\phi^s = \sum_{l=0}^{\infty} A_l h_l^{(1)}(k_p r)P_l(\cos\theta) \tag{3.62}$$

$$\chi^s = \sum_{l=0}^{\infty} B_l h_l^{(1)}(k_s r)P_l(\cos\theta), \tag{3.63}$$

whereas standing waves are set up by the fields refracted into the sphere:

$$\phi^r = \sum_{l=0}^{\infty} C_l j_l(k'_p r)P_l(\cos\theta) \tag{3.64}$$

$$\chi^r = \sum_{l=0}^{\infty} D_l j_l(k'_s r)P_l(\cos\theta). \tag{3.65}$$

The incoming wave may be assumed, without loss of generality, to be propagating along the z axis. It is expanded into spherical Bessel functions as

$$\phi^i = \phi_0 e^{ik_p r\cos\theta} = \phi_0 \Sigma_{l=0}^{\infty}(2l+1)i^l j_l(k_p r)P_l(\cos\theta). \tag{3.66}$$

To solve for the coefficients A_l, B_l, C_l and D_l requires the four continuity conditions on displacements and stresses at the surface of the sphere, at radius a:

$$u_r^i + u_r^s = u_r^r \tag{3.67}$$

$$u_\theta^i + u_\theta^s = u_\theta^r \tag{3.68}$$

$$\tau_{rr}^{i} + \tau_{rr}^{s} = \tau_{rr}^{r} \tag{3.69}$$

$$\tau_{r\theta}^{i} + \tau_{r\theta}^{s} = \tau_{r\theta}^{r}. \tag{3.70}$$

After a great deal of algebra, we arrive at the following matrix equation for the coefficients, in which we have taken advantage of the orthogonality property of the spherical harmonics to factorize the matrix into blocks, each for a single value of l:

$$\begin{Bmatrix} M_{11} & M_{12} & M_{13} & M_{14} \\ M_{21} & M_{22} & M_{23} & M_{24} \\ M_{31} & M_{32} & M_{33} & M_{34} \\ M_{41} & M_{42} & M_{43} & M_{44} \end{Bmatrix} \begin{Bmatrix} A_l \\ B_l \\ C_l \\ D_l \end{Bmatrix} = \phi_0 \begin{Bmatrix} V_1 \\ V_2 \\ V_3 \\ V_4 \end{Bmatrix} \tag{3.71}$$

The explicit forms of the matrix elements are given below.

$$M_{11} = nh_n^{(1)}(k_p a) - k_p a h_{n+1}^{(1)}(k_p a) \tag{3.72}$$

$$M_{12} = -n(n+1)h_n^{(1)}(k_s a) \tag{3.73}$$

$$M_{13} = nj_n(k'_p a) - k'_p a j_{n+1}(k'_p a) \tag{3.74}$$

$$M_{14} = -n(n+1)j_n(k'_s a) \tag{3.75}$$

$$M_{21} = h_n^{(1)}(k_p a) \tag{3.76}$$

$$M_{22} = -(n+1)h_n^{(1)}(k_s a) + k_s a h_{n+1}^{(1)}(k_s a) \tag{3.77}$$

$$M_{23} = j_n(k'_p a) \tag{3.78}$$

$$M_{24} = -(n+1)j_n(k'_s a) + k'_s a j_{n+1}(k'_s a) \tag{3.79}$$

$$M_{31} = (n^2 - n - \tfrac{1}{2}k_s^2 a^2)h_n(k_p a) + 2k_p a h_{n+1}^{(1)}(k_p a) \tag{3.80}$$

$$M_{32} = -n(n+1)\{(n-1)h_n^{(1)}(k_s a) - k_s a h_{n+1}^{(1)}(k_s a)\} \tag{3.81}$$

$$M_{33} = (\mu'/\mu)\{(n^2 - n - \tfrac{1}{2}k'^2_s a^2)j_n(k'_p a) + 2k'_p a j_{n+1}(k'_p a)\} \tag{3.82}$$

$$M_{34} = -(\mu'/\mu)n(n+1)\{(n-1)j_n(k'_s a) - k'_s a j_{n+1}(k'_s a)\} \tag{3.83}$$

$$M_{41} = (n-1)h_n^{(1)}(k_p a) - k_p a h_{n+1}^{(1)}(k_p a) \tag{3.84}$$

$$M_{42} = -(n^2 - 1 - \tfrac{1}{2}k_s^2 a^2)h_n^{(1)}(k_s a) - k_s a h_{n+1}^{(1)}(k_s a) \tag{3.85}$$

$$M_{43} = (\mu'/\mu)\{(n-1)j_n(k'_p a) - k'_p a j_{n+1}(k'_p a)\} \tag{3.86}$$

$$M_{44} = -(\mu'/\mu)\{(n^2 - 1 - \tfrac{1}{2}k_s^2 a^2)j_n(k'_s a) + k'_s a j_{n+1}(k'_s a)\} \tag{3.87}$$

$$V_1 = -i^n(2n+1)\{nj_n(k_p a) - k_p a j_{n+1}(k_p a)\} \tag{3.88}$$

$$V_2 = -i^n(2n+1)\{j_n(k_pa)\} \tag{3.89}$$

$$V_3 = -i^n(2n+1)\left\{(n^2-n-\tfrac{1}{2}k_s^2a^2)j_n(k_pa) + 2k_paj_{n+1}(k_pa)\right\} \tag{3.90}$$

$$V_4 = -i^n(2n+1)\left\{(n-1)j_n(k_pa) - k_paj_{n+1}(k_pa)\right\}. \tag{3.91}$$

The above equations may be solved numerically for general cases, but analytical solutions are quite easy to extract for very large and very small scatterers. The case of small scatterers is of special interest in nondestructive testing applications, as inclusions of slag and similar metallurgical features are often small compared with the ultrasonic wavelength. It is interesting to see how strongly such features are likely to scatter.

For small arguments, the spherical Bessel functions may be expanded in ascending powers of the argument, with leading terms which are given in equations (3.60) and (3.61). When this is done, the expressions simplify to give the following matrix equation for the lowest order coefficients:

$$\begin{Bmatrix} -\tfrac{1}{3}k_p^2a^2 & 0 & -\tfrac{1}{3}k'^2_pa^2(\mu'/\mu) & 0 \\ 1 & 0 & (\mu'/\mu) & (\mu'/\mu) \\ k_p^2a^2(\tfrac{2}{3}-\tfrac{1}{2}k_s^2/k_p^2) & 0 & k'^2_pa^2(\tfrac{2}{3}-\tfrac{1}{2}k'^2_s/k'^2_p) & 0 \\ 1 & 1/(2k_sr) & -1 & 1 \end{Bmatrix} \times \begin{Bmatrix} A_0 \\ B_0 \\ C_0 \\ D_0 \end{Bmatrix} = \frac{\phi_0}{k_pa}\begin{Bmatrix} -1 \\ 1 \\ 2 \\ -2 \end{Bmatrix} \tag{3.92}$$

from which one obtains

$$A_0 = \tfrac{1}{3}(k_pa)^3\left[1 - \frac{3(k_s/k_p)^2}{3(\mu/\mu')(k_s/k_p)^2 - 4((\mu/\mu') - 1)}\right]. \tag{3.93}$$

This shows the characteristic variation with the cube of the radius of the scatterer in the small-scatterer (Rayleigh) limit. Further terms in the expansion have coefficients

$$A_1 = \frac{\rho - \rho'}{3\rho}(k_pa)^3 \tag{3.94}$$

$$B_1 = -(k_s^2/k_p^2)A_1 \tag{3.95}$$

and we shall compare these results below with the approximate solutions of the integral equations for scattering.

3.1.3 Scattering of elastic waves by a cylinder

There are many situations (for example, holes for bolts or rivets or channels for cooling water) in which structures contain cylindrical cavities. For incident waves perpendicular to the cylindrical axis these cases may be treated in two dimensions. The numerical modelling techniques described later in this chapter are also limited to two dimensions. For comparisons of analytical and numerical results in the next chapter, we develop here the expressions for the scattering by a circular cylindrical cavity.

White (1958) examined the scattering in this situation, both theoretically and experimentally. His theoretical work was repeated, with minor corrections, by Lewis, Kraft and Horn (1976). We shall take advantage of the simplifications which appear when the incident wavevector is perpendicular to the axis of the cylinder, and when the displacements are also perpendicular to that axis. If we take the axis of the cylinder to be the z axis, we then have

$$u_z = 0 \qquad \tau_{zi} = 0 \quad \forall i \qquad \frac{\partial}{\partial z} \to 0. \tag{3.96}$$

The wave equation in polar coordinates for a frequency ω is

$$\frac{1}{r}\frac{\partial}{\partial r}\left(r\frac{\partial \phi}{\partial r}\right) + \frac{1}{r^2}\frac{\partial^2 \phi}{\partial \theta^2} + \frac{\omega^2}{c^2}\phi = 0, \tag{3.97}$$

which may be separated by taking

$$\phi(r,\theta) = R(r)\Theta(\theta) \tag{3.98}$$

to give

$$r^2 R'' + rR' + \left(\frac{\omega^2}{c^2}r^2 - n^2\right)R = 0 \tag{3.99}$$

and

$$\Theta'' + n^2\Theta = 0. \tag{3.100}$$

Here n must be an integer, in order that the solutions Θ may be single-valued functions of θ:

$$\Theta(\theta) = e^{\pm in\theta}, \tag{3.101}$$

and then the solutions of (3.99) are cylindrical Bessel functions of order n. As before, an incident plane wave may be expanded in terms of Bessel functions of the first kind:

$$\phi(x) = \phi_0 \sum_{n=0}^{\infty} \varepsilon_n i^n J_n(kr)\cos n\theta \tag{3.102}$$

where

$$\varepsilon_n = \begin{cases} 1 & n=0 \\ 2 & n\geqslant 1. \end{cases} \tag{3.103}$$

Similarly, an outgoing wave is represented by a sum of Hankel functions of the first kind. The differential operators in cylindrical coordinates give the displacements in terms of the potentials ϕ and ψ:

$$u_r = \frac{\partial \phi}{\partial r} + \frac{1}{r}\frac{\partial \psi}{\partial r} \tag{3.104}$$

$$u_\theta = \frac{1}{r}\frac{\partial \phi}{\partial \theta} - \frac{\partial \psi}{\partial r} \tag{3.105}$$

$$\tau_{rr} = \lambda \nabla^2 \phi + 2\mu\left[\frac{\partial^2 \phi}{\partial r^2}\frac{\partial}{\partial r}\left(\frac{1}{r}\frac{\partial \psi}{\partial \theta}\right)\right] \tag{3.106}$$

$$\tau_{\theta\theta} = \lambda \nabla^2 \phi + 2\mu\left[\frac{1}{r}\left(\frac{\partial \phi}{\partial r} + \frac{1}{r}\frac{\partial^2 \phi}{\partial \theta^2} + \frac{1}{r}\frac{\partial \psi}{\partial \theta} - \frac{\partial^2 \psi}{\partial r \partial \theta}\right)\right] \tag{3.107}$$

$$\tau_{r\theta} = 2\mu\left[\frac{1}{r}\frac{\partial^2 \phi}{\partial r \partial \theta} - \frac{1}{r^2}\frac{\partial \phi}{\partial \theta}\right] + \mu\left[\frac{1}{r^2}\frac{\partial^2 \psi}{\partial \theta^2} - r\frac{\partial}{\partial r}\left(\frac{1}{r}\frac{\partial \psi}{\partial r}\right)\right]. \tag{3.108}$$

The scattered wave is expanded as

$$\phi^s(r,\theta) = \sum_{n=0}^{\infty} A_n H_n^{(1)}(k_p r)\cos n\theta \tag{3.109}$$

$$\psi^s(r,\theta) = \sum_{n=0}^{\infty} B_n H_n^{(1)}(k_s r)\sin n\theta. \tag{3.110}$$

Substituting these expansions into the equations for the stresses, and setting the surface tractions on the cavity to zero, we arrive at the following matrix equation for the coefficients. Again we have taken advantage of the orthogonality property of the angular functions (sines and cosines of different multiples of angle) to factorize the matrix into blocks, each for a single value of n.

$$\begin{Bmatrix} M_{11} & M_{12} \\ M_{21} & M_{22} \end{Bmatrix}\begin{Bmatrix} A_n \\ B_n \end{Bmatrix} = \phi_0 \begin{Bmatrix} V_1 \\ V_2 \end{Bmatrix} \tag{3.111}$$

The explicit forms of the matrix elements are:

$$M_{11} = (n^2 + n - \tfrac{1}{2}k_s^2 a^2)H_n^{(1)}(k_p a) - k_p a H_{n-1}^{(1)}(k_p a) \tag{3.112}$$

$$M_{12} = n(n+1)H_n^{(1)}(k_s a) - nk_s a H_{n-1}^{(1)}(k_s a) \tag{3.113}$$

$$M_{21} = n(n+1)H_n^{(1)}(k_p a) - nk_p a H_{n-1}^{(1)}(k_p a) \tag{3.114}$$

$$M_{22} = -(n^2+n-\tfrac{1}{2}k_s^2a^2)H_n^{(1)}(k_sa) + k_saH_{n-1}^{(1)}(k_sa) \qquad (3.115)$$

$$V_1 = -\varepsilon_n i^n\left\{(n^2+n-\tfrac{1}{2}k_s^2a^2)J_n(k_pa) - k_paJ_{n-1}(k_pa)\right\} \qquad (3.116)$$

$$V_2 = -\varepsilon_n i^n\left\{n(n+1)H_n^{(1)}(k_pa) - k_paJ_{n-1}(k_pa)\right\}. \qquad (3.117)$$

The quantities which will be of interest are the scattering cross-sections, which may be defined in terms of the coefficients A_n and B_n as

$$\sigma_{pp} = \frac{2}{k_pa}\sum_{n=0}^{\infty}\varepsilon_n|A_n|^2 \qquad (3.118)$$

and

$$\sigma_{ps} = \frac{2}{k_pa}\sum_{n=0}^{\infty}\varepsilon_n|B_n|^2, \qquad (3.119)$$

which are the cross-sections for scattering incident compression waves into compression and shear waves respectively. These quantities have been computed by White (1958) and by Lewis *et al* (1976) for various values of k_pa.

Figure 24

The cross-section for scattering a compression wave into compression (σ_{pp}) and shear waves (σ_{ps}) by a cylinder, in a medium with $c_p/c_s = 1.45$ (after White 1958).

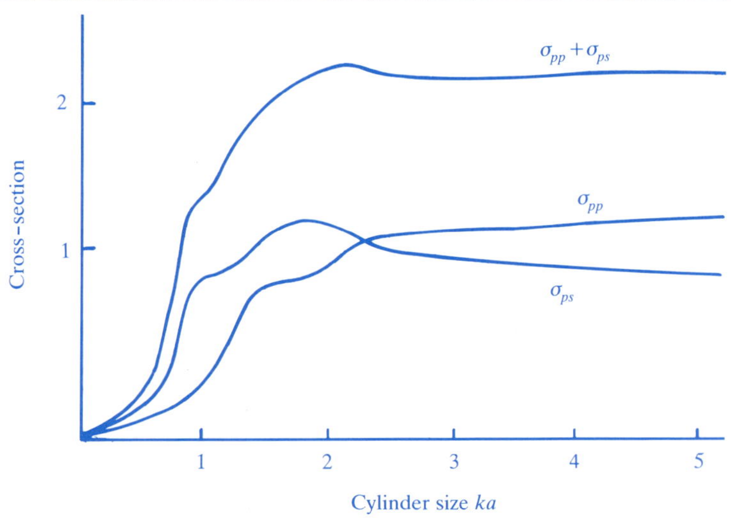

Typical results are shown in Figure 24, where an initial rise proportional to $(k_pa)^3$ is clearly visible, followed by some oscillations, until the cross-section becomes nearly constant for ka a few times greater than unity. This suggests that there is only a fairly narrow range of $k_p a$, say $\frac{1}{3}<k_p a<3$, within which detailed calculations are necessary. For smaller scatterers the Rayleigh limit applies, and for larger scatterers a purely geometrical treatment with a constant cross-section may be used.

Figure 25

The cross-section for scattering a shear wave into compression (σ_{sp}) and shear (σ_{ss}) waves by a cylinder, in a medium with $c_p/c_s = 1.45$ (after White 1960).

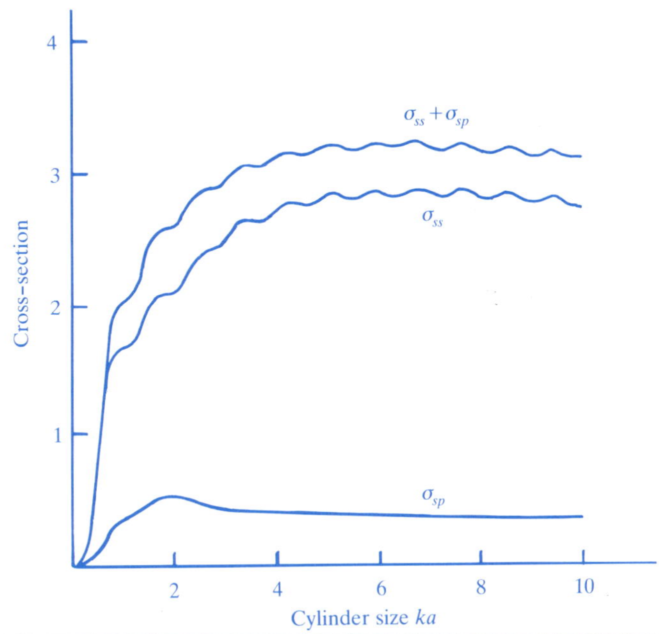

For shear waves, however, Figure 25 shows that although the same qualitative behaviour is seen at small $k_s a$, the oscillatory behaviour persists to much larger values of $k_s a$, where it is found in the shear–to–shear scattering cross-section.

Oscillatory behaviour of this sort often arises from resonant behaviour, so it is interesting to ask what are the normal modes of the cylindrical cavity. That is, we seek solutions to (3.111) when there is no incident wave. For non-trivial solutions the determinant of the matrix **M** must be zero. This problem has analogues in acoustics, most notably the normal modes of a fluid in a cylindrical cavity, where the solutions include the Whispering Gallery modes investigated by Rayleigh (1910) following his observations in 1880 inside the dome of St Paul's Cathedral. For a cylindrical cavity in a solid, Pao and Mow (1976) have evaluated the resonant frequencies, and Überall (1978) has reviewed the work which shows how the scattering cross-sections may be written in terms of these resonances. The essential point is that if the scattered field is expanded, as above, in eigenfunctions then the coefficients A_n and B_n may be written as

$$A_n = e^{2i\xi_n}\left[\sum_{s=1}^{\infty}\frac{M_n^s}{k^p a-\alpha_n^s+\frac{i}{2}\Gamma_n^s}+2ie^{-i\xi_n}\sin\xi_n\right] \tag{3.120}$$

$$B_n = e^{2i\xi'_n}\left[\sum_{s=1}^{\infty}\frac{M'^s_n}{k^p a-\alpha_n^s+\frac{i}{2}\Gamma_n^s}+1\right]. \tag{3.121}$$

Here the quantities α_n^s are the sth roots of the determinant of $\mathbf{M}_n$, corresponding to resonances in which the vibrations have n nodal planes ($\cos n\theta$ or $\sin n\theta$ angular behaviour). The phase shifts involved in the scattering from an empty cylinder are ξ and ξ' for compression and shear outgoing waves. Γ_n^s, the width of the resonance, and M_n^s, the strength of the resonance, are described in detail in the works cited above. For practical applications, it should be noted that the expressions involve solutions of equations involving Bessel functions, and that no economy of computational effort arises from writing the scattering in terms of the resonances. This formulation is useful, however, in showing the physical origin of the resonant behaviour in the scattering cross-sections. It also shows that the scattering by a cylinder will generally resemble the scattering by an impenetrable (evacuated) cylinder (the second term in each bracket in the scattering cross-section) except near a resonance, where the first term will dominate.

The resonances are of several types. There are Franz waves, which are the analogues on curved surfaces of Stoneley waves on flat surfaces. They differ from Stoneley waves in two respects: they are attenuated because energy is radiated tangentially away from the cylinder, and there are many modes of slightly different speeds, one for each circumferential mode n. There is a single Rayleigh mode, which will in general have a complex frequency, which corresponds to the leaky Rayleigh wave at a plane interface. There are also the whispering gallery modes. For large n these correspond to compression and shear waves circling round the surface of the cylinder, although they will also leak into the bulk. As the radius of the cylinder becomes large, the whispering gallery modes degenerate into the lateral waves mentioned in connection with buried sources in Chapter 2. The relationship between creeping waves and lateral waves was demonstrated by Rulf (1967) for scalar waves in two dimensions, although as pointed out by Überall (1973) his procedure could be improved by explicit inclusion of surface wave terms.

In the limit of cylinders whose radii are large compared with the wavelength, all the modes of the cylinder may be equated with corresponding modes of a plane surface. An analysis of the asymptotic forms of the Bessel functions for large values of the order n and argument ka shows that the errors made in treating the surfaces as plane are of order $1/(ka)^2$ times the terms which are kept. For a typical ultrasonic experiment, with a wavelength of the order of a millimetre, the modes propagating in a pipe wall may be treated as bulk waves, Rayleigh surface waves and lateral waves provided that the pipe radius a is five centimetres or more. The resonance analysis of Dickey, Frisk and Überall (1976) of the modes of an elastic cylinder shows the convergence of the poles onto those for a plane surface, and confirms that the plane surface approximation is accurate for ka greater than 100π. Rulf (1969) also gives results for the Rayleigh wave speed as a function of the radius of the cylinder. This convergence is rather slow, and causes problems in the analytic treatment of the scattering by cylinders which are large compared to the wavelength. Brind, Achenbach and Gubernatis (1984) have studied this situation in detail, and find that good results can be obtained with a Kirchhoff method for back scattering, but that a ray theory analysis which takes account of the waves which creep around the surface is necessary for forward scattering. An improved convergence of the partial wave series may be gained by using the Sommerfeld–Watson transformation (Doolittle, Überall and Ugincius 1968). The poles which occur in the inversion can be interpreted in terms of repeated circumnavigations of the cylinder by various types of creeping wave. Similar analysis is possible for the scattering from spheres at high frequencies (Williams and Marston 1985, 1986). Again, there is a clear correspondence with ray theory. The Kirchhoff and ray methods are described in Sections 3.2.2 and 3.3 below.

3.2 Scattering in the integral equation formalism

The integral form of the wave equation was derived in the previous chapter. Here we consider how the equation may be manipulated to give an integral equation for the scattering of an elastic wave from a defect. For simplicity, we treat waves of a single frequency ω, and it should be recalled that if the scattering of a pulse is required an additional integral, inverting the Fourier transform, will be required. This means, in turn, that the integral equation which contains ω as a parameter must be solved for all values of ω.

When the volume V in equation (2.255) contains a defect, the elastic properties of the material vary inside V. The variation may be described by defining V to be the volume *outside* the flaw, V' the volume *inside*, when

$$c_{ijkl}\rightarrow c_{ijkl}+\delta c_{ijkl}H(\mathbf{r}) \tag{3.122}$$

and

$$\rho \rightarrow \rho + \delta\rho H(\mathbf{r}) \tag{3.123}$$

where $H(\mathbf{r})$ is defined to be unity inside V' and zero outside, and the differences in material properties inside and outside the flaw are

$$\delta c_{ijkl} = c_{ijkl}' - c_{ijkl} \tag{3.124}$$

$$\delta\rho = \rho' - \rho, \tag{3.125}$$

the primed values being those of the material in the flaw. The elastic wave equation may then be written as

$$\begin{aligned} c_{ijkl} u_{k,lj} + \rho\omega^2 u_i \\ = -(\delta\rho\, \omega^2 u_i H + \delta c_{ijkl} u_{k,jl} H + \delta c_{ijkl} u_{k,l} H_{,j}) \end{aligned} \tag{3.126}$$

where $H_{,j}$ is the derivative of the step function H with respect to r_j, which may be written in terms of a delta function

$$H_{,j}(\mathbf{r}) = -\hat{n}_j \delta(\mathbf{r} - \mathbf{r}^{S'}), \tag{3.127}$$

$\mathbf{r}^{S'}$ being on the surface of the flaw, and $\hat{\mathbf{n}}$ being the unit vector normal to the surface S' pointing out of V'. The right-hand side of equation (3.126) may be treated as the body force f_i in equation (2.255), and inside the entire region of interest $V - V'$ we may write the total wavefield $\mathbf{u}$ as the sum of the incident and scattered fields:

$$\mathbf{u} = \mathbf{u}^i + \mathbf{u}^s \tag{3.128}$$

so that the scattering equation becomes

$$\begin{aligned} u_k(\mathbf{r}) = u_k^i(\mathbf{r}) \\ + \int_{V'} (\delta\rho\, \omega^2 G_{ki}(\mathbf{r}|\mathbf{r}') u_i(\mathbf{r}') - \delta c_{ijmn} u_{m,n}(\mathbf{r}') G_{ki,j}(\mathbf{r}|\mathbf{r}')) dV'. \end{aligned} \tag{3.129}$$

This is an integral equation representing the scattering of the incident wave $\mathbf{u}^i$. Note that there is a notional incident field inside the scatterer, which is equal to the field that would have been present if the scatterer were absent.

For many purposes it is more convenient to obtain a surface integral formulation of the scattering. The starting point is again equation (2.255), but instead of integrating throughout the entire volume we confine the integral to the region outside the scatterer. This volume has two bounding surfaces, the surface of the scatterer and an enclosing surface at infinity. We assume that the scattered field is zero on the enclosing surface. It then follows that the scattered wave outside the scattering surface is given by

$$u_k^s(\mathbf{r}) = \int_{S'} (u_i(\mathbf{r}') \Sigma_{ijk}(\mathbf{r}|\mathbf{r}') \hat{n}_j - \tau_{ij}(\mathbf{r}') G_{ik}(\mathbf{r}|\mathbf{r}') \hat{n}_j) dS'. \tag{3.130}$$

Note that because we are outside the scatterer here, $\hat{\mathbf{n}}$ is a unit vector pointing from the surface S' into the medium supporting the wave. This is simply an expression of the Huygens–Kirchhoff principle (Baker and Copson 1950), which was derived for elastic waves by de Hoop (1958), Kupradze (1963) and Pao and Varatharajulu (1976), among others. Again it is an integral equation for the scattered field. Kupradze's formulation was rather more general than the version we shall present here, and involves an arbitrary constant. Gniadek and Petykiewicz (1971) have given a careful analysis of the more general form, and have shown how different choices of the arbitrary constant correspond to different discontinuities at a diffracting edge. They relate the general theory to the edge integrals added to his theory by Knopoff (1956): Knopoff worked by analogy with electromagnetic diffraction theory and did not justify the extra edge integrals in the elastodynamic case. Gniadek and Petykiewicz (1971) also show how the Kirchhoff theory may be manipulated to exhibit the form of a geometrical elastodynamic field with an edge diffraction correction, in the spirit of Keller's theory of diffraction. As we know that our incident wave satisfies the homogeneous wave equation, we can also write equation (3.130) in the form

$$u_k^s(\mathbf{r}) = \int_{S'} (u_i^s(\mathbf{r}')\Sigma_{ijk}(\mathbf{r}|\mathbf{r}')\hat{n}_j - \tau_{ij}^s(\mathbf{r}')G_{ik}(\mathbf{r}|\mathbf{r}')\hat{n}_j)dS'. \qquad (3.131)$$

Alternative derivations of this equation are possible: for example, it follows directly from the reciprocity relation, or from the Helmholtz interior and exterior formulae (Pao and Mow 1973, page 140). Unlike equation (3.130), this does not involve any assumptions about the behaviour of the scattered field at infinity. Equation (3.131), despite its appearance, is **not** an integral equation for the scattered field. This is obvious, as it would appear to lead to a scattered field which is independent of the incident field. The point is that the boundary conditions on the surface S' cannot be included in equation (3.131), as the integral involves only part of the wavefield: they can be included in equation (3.130), which involves the total wavefield. Another way of putting this is that the expression of Huygens' principle in the form of equation (3.131), although correct, gives no guidance as to the strengths of the effective sources at the surface which give rise to the scattered waves.

The earliest applications of the Kirchhoff approximation were to diffraction through an aperture, when it was assumed that the screen surrounding the aperture was perfectly absorbing. The effect of this assumption is that no reflected waves are found. In applications to scattering it is more common to assume that the values and gradients of the scattered field may be determined from the incident field, and that the field is non-zero only in regions which would be insonified in a geometric ray picture. That is, there is a sharply defined shadow zone in which the incident waves are assumed to be zero. The scattered field is then calculated by performing the integral in equation (3.131) with the assumed field at the surface of the scatterer. This modification of the Kirchhoff model reinstates the reflected waves (de Hoop 1958; Bouwkamp 1954), but the head waves cannot be recovered. Despite these drawbacks, the Kirchhoff approximation has been quite extensively applied to the scattering of ultrasound (for example Neubauer 1963; Johnson 1976; Haines and Langston 1980). Ogilvy (1987) has used the variational principle to assess the accuracy of the approximation. The method is not restricted to cracks, and the basic assumption for arbitrary shapes is that the scattered field and its gradients at the surface of the scatterer are equal to those of the incident field, multiplied by a reflection coeffieient, in the region which is reached by the incident wave (the insonified region), and that the total field is zero in the shadow zone. In view of its wide application, we shall discuss the Kirchhoff approximation in more detail below.

Approximate solutions to the integral equations may be sought in several other ways, but the ones which have been most commonly used are the Born and quasistatic approximations. The Born approximation is suited to weak scattering, and consists of approximating the displacement $\mathbf{u}$ in the interior of the flaw by the incident field $\mathbf{u}^i$. In principle, this is the first term of an expansion, and a better approximation could be obtained by substituting the resulting total field back into the integral: in practice the expansion is usually halted at the first term. For stronger scattering, a better estimate of the field inside the flaw is obtained by using solutions of the corresponding static problem: this technique is suitable when the wavelength of the elastic waves is large compared with the flaw, and constitutes the quasistatic approximation.

Direct solutions of the integral equations have been attempted by three methods: finite difference; finite element; and moments. The finite difference approach is more familiar in its application to differential equations, and this will be described later in this chapter. For integral equations great care is necessary, as the Green function has a singularity at $\mathbf{r} = \mathbf{r}'$, and the contribution from the singularity often dominates the solution. Banaugh and Goldsmith (1963) were successful in applying this method to the scattering of acoustic waves from cylinders with various cross-sections. Another problem with the integral equation method, when formulated as a boundary integral equation, is that the solution is not unique at wavevectors corresponding to eigenstates of the scattering region. Various methods exist (Burton and Miller 1971) for overcoming this difficulty. More recently, Reut (1982, 1985) has proposed a modification of the Helmholtz integral equation method which replaces the integration of the singular terms over flat rectangular surface elements by analytic results, leaving only simple contour integrals to be evaluated numerically. Tobocman (1984a,b) also advocates the use of the integral equation method for acoustic wave scattering, but uses instead the combined Helmholtz integral equation method of Schenck (1968) to ensure uniqueness. In this scheme, the surface points are augmented by a set of interior points, and the displacements at those points incorporated in the equations by using the interior Helmholtz formula. The resulting overdetermined set of equations is solved by a least squares procedure. Finite element methods are also described later in this chapter, and the singular behaviour of the kernel must again be handled. The method of moments involves an expansion in basis functions. It was applied by Tan (1975, 1976, 1977a,b) to the scattering of elastic waves by a ribbon-like defect. These calculations were for two-dimensional systems. The three-dimensional system which is most commonly treated is the scattering by a penny-shaped crack, and the integral equation for this problem has been treated by several authors, including Mal (1968, 1970) and Sih and Loeber (1969). Their work is reviewed by Kraut (1976) and Datta (1978). More recently the problem has been revisited by Martin (1981) and reviewed by Martin and Wickham (1983).

In cases where the analytic form of the Green function is known, as in a homogeneous region, the boundary integral equation method can be applied. Here the limit of the integral equation as the field point $\mathbf{r}$ approaches the boundary is taken, with the improper integrals which result being interpreted in terms of their limiting values. Essentially, the Green function is used to describe how an element of the boundary will interact with another element of the boundary. The method only requires the boundary to be discretized, as the wavefield at other points can be obtained from the Green function. The method appears attractive because the dimensionality of the boundary is one less than that of the region it encloses, so a smaller number of points should be necessary in the discretization. However, each point on the boundary is linked through the Green function to every other point, whereas in a finite element model each volume element is linked only to its immediate neighbours. When the resulting matrix equations are set up, this means that the boundary element method leads to a small, full matrix whereas the finite element method gives a larger, banded matrix. To take a specific example, consider a problem on a square region. If each side is divided into n elements, the boundary integral method will give a $4n \times 4n$ symmetric matrix, which can be inverted in a time proportional to $\frac{1}{6}(4n)^3$. The equivalent finite element problem is $n^2 \times n^2$, with a band width of about n, leading to a solution time proportional to $\frac{1}{2}n^4$. Thus the boundary element method only becomes advantageous for $n \gtrsim 20$. In addition, the integrals involved are singular where they contain $G_{ij}(\mathbf{r}|\mathbf{r}')$ for $\mathbf{r} = \mathbf{r}'$. A further problem arises if numerical values of the field are required at many points remote from the boundary, as the values must be obtained by numerical integration using the boundary values and the Green's functions.

We shall not discuss the boundary integral equation method in detail here, but note that there are three main approaches. The first is a direct time-domain method, as pioneered by Friedman and Shaw (1962): this uses a retarded potential scheme. The other approaches use integral transforms in the time domain. The Laplace transform was used by the elastodynamic boundary integral method was investigated by Cruse and Rizzo (1968), Cruse (1968), and Manolis and Beskos (1980, 1981). Fourier transforms have been used extensively by Niwa and co-workers (see Niwa, Kitahara and Ikeda 1984) and by Kobayashi and Nishimura (1982). Manolis (1983) has compared the direct and the two transform methods. According to Tsao, Varadan and Varadan (1982) the method suffers from spurious resonances in the scattered field at frequencies which correspond to interior resonances. The t-matrix method described below does not have this drawback. Kobayashi and Nishimura (1982) describe techniques for avoiding fictitious eigenfrequencies when fast Fourier transform methods are used, although simple interpolation (Niwa, Hirose and Kitahara 1986) appears to be effective.

3.2.1 General properties of elastic wave scattering

Some important general results in elastic wave scattering were derived by Gubernatis, Domany and Krumhansl (1977) from the integral equation. Their results illustrate some of the important new effects which occur in the scattering of elastic waves compared with scalar (acoustic) waves or even purely transverse vector (light) waves. To calculate the differential scattering cross-section, as defined in Chapter 2, it is necessary to evaluate the scattered wave far from the flaw. This requires the asymptotic form of the Green function. As shown in equation (2.266), $\mathbf{G}(\mathbf{r}|\mathbf{r}')$ depends only on the relative vector $\mathbf{R} = \mathbf{r}-\mathbf{r}'$, where $\mathbf{r}$ is the vector to the point of observation and $\mathbf{r}'$ is a vector to a point inside the scatterer. One can immediately derive the limits for large $r = |\mathbf{r}|$:

$$R^{-1} \sim r^{-1} \tag{3.132}$$

$$R \sim r - \mathbf{r}' \cdot \mathbf{r} \tag{3.133}$$

$$\frac{\partial}{\partial r_i} \frac{e^{ikR}}{R} \sim ik\hat{r}_i e^{-i\mathbf{k}\cdot\mathbf{r}'} \frac{e^{ikr}}{r} \tag{3.134}$$

with higher derivatives giving expressions similar to equation (3.134) including a factor $(ik)\hat{r}_i$ for each derivative. The far-field scattered wave then has a displacement

$$u_i^s \sim \hat{r}_i \hat{r}_j f_j(k_p \hat{\mathbf{r}}) \frac{e^{ik_p r}}{k_p r} + (\delta_{ij} - \hat{r}_i \hat{r}_j) f_j(k_s \hat{\mathbf{r}}) \frac{e^{ik_s r}}{k_s r}, \tag{3.135}$$

where the scattering vector $\mathbf{f}(\mathbf{k})$ is defined by

$$f_i(\mathbf{k}) = \frac{k^3}{4\pi\rho\omega^2}\left\{\delta\rho\omega^2 \int_{V'} u_i e^{-i\mathbf{k}\cdot\mathbf{r}'}\, dV' + ik\hat{r}_j \delta c_{ijkl} \int_{V'} e_{kl} e^{-i\mathbf{k}\cdot\mathbf{r}'}\, dV'\right\}. \quad (3.136)$$

Although the symbol $\mathbf{f}$ is being used here for the scattering vector, there should be no confusion with its use elsewhere to represent a body force. Note that the definition of $\mathbf{f}$ used here differs from that of Gubernatis, Domany and Krumhansl (1977) because they take the fundamental outgoing wave to have the form $\exp(ikr)/r$ whereas we use $\exp(ikr)/(kr)$. The present $\mathbf{f}$ vector therefore has the dimensions of length.

The scattered field can obviously be separated into longitudinal and transverse components,

$$u_i^s \sim \mathrm{P}_i \frac{e^{ik_p r}}{k_p r} + S_i \frac{e^{ik_s r}}{k_s r}, \quad (3.137)$$

with the compression and shear wave scattering amplitudes

$$\mathrm{P}_i(\theta,\phi) = \hat{r}_i \hat{r}_j f_j(k_p \hat{\mathbf{r}}) \quad (3.138)$$

and

$$\mathrm{P}_i(\theta,\phi) = (\delta_{ij} - \hat{r}_i \hat{r}_j) f_j(k_s \hat{\mathbf{r}}). \quad (3.139)$$

To derive the scattering cross-section, it is necessary to evaluate the stress arising from the scattered wave. For an isotropic medium this may be expressed in terms of the usual constants λ and μ as

$$\tau_{ij}^s \sim i\lambda k_p \mathrm{P}_k \hat{r}_k \delta_{ij} \frac{e^{ik_p r}}{k_p r} + i\mu k_p (P_i \hat{r}_j + \mathrm{P}_j \hat{r}_i) \frac{e^{ik_p r}}{k_p r} + i\mu k_s (S_i \hat{r}_j + S_j \hat{r}_i) \frac{e^{ik_p r}}{k_s r} \cdot \quad (3.140)$$

Then, if the incident wave is a mixture of compression and shear waves travelling parallel to the z axis,

$$u_i^i = p_i e^{i(k_p z - \omega t)} + s_i e^{i(k_s z - \omega t)}, \quad (3.141)$$

which has amplitudes $|\mathbf{p}|$ for the compressional component and $|\mathbf{s}|$ for the shear component, the incident intensity is

$$F = \tfrac{1}{2}\omega\left[k_p(\lambda + 2\mu)|\mathbf{p}|^2 + k_s \mu |\mathbf{s}|^2\right] \quad (3.142)$$

and the limiting value of the scattered power as $r \to \infty$ is

$$\left\langle r^2 \hat{r}_i \tau_{ij}^s u_{j,t}^s \right\rangle \sim \tfrac{1}{2}\omega\left[k_p(\lambda + 2\mu)|\mathrm{P}|^2 + k_s \mu |\mathrm{S}|^2\right] \cdot \quad (3.143)$$

In equations (3.142) and (3.143) the appropriate modifications must be made when complex field notation is being used, as described in Chapter 2. The differential scattering cross-section is then

$$\frac{d\sigma(\omega)}{d\Omega} = \frac{k_p(\lambda+2\mu)|\mathrm{P}|^2+k_s\mu|\mathbf{S}|^2}{k_p(\lambda+2\mu)|\mathbf{p}|^2+k_s\mu|\mathbf{s}|^2}. \tag{3.144}$$

Equation (3.144) shows very clearly that mode conversion generally accompanies scattering. For if the incident wave is purely compressional and of unit amplitude the differential cross-section is

$$\frac{d\sigma(\omega)}{d\Omega} = |\mathrm{P}|^2+\frac{k_p}{k_s}|\mathrm{S}|^2, \tag{3.145}$$

whereas, if the incident wave is purely transverse and of unit amplitude,

$$\frac{d\sigma(\omega)}{d\Omega} = \frac{k_s}{k_p}|\mathrm{P}|^2+|\mathbf{S}|^2. \tag{3.146}$$

This clearly shows that the form of the cross-section depends on the type of the incident wave, but always contains contributions from both types of scattered wave.

Another important result derived by Gubernatis, Domany and Krumhansl (1977) is the elastodynamic equivalent of the optical theorem. For scalar waves, one has the simple relationship that the total cross section is the total solid angle (4π) times the forward scattering amplitude :

$$\sigma(\omega) = 4\pi f(\theta=0). \tag{3.147}$$

For elastic waves, the corresponding formula is more complicated. The total displacement field is the sum of the incident and scattered fields, so that the intensity is

$$F_i = \tau^i_{ij}u^i_{j,t}+(\tau^i_{ij}u^s_{j,t}+\tau^s_{ij}u^i_{j,t})+\tau^s_{ij}u^s_{j,t} \tag{3.148}$$

which may be written as the sum of incident, interference and scattered intensities:

$$F_i = F^i_i+F^{int}_i+F^s_i. \tag{3.149}$$

Now, as $\mathbf{F}^i$ is the power being put into the system, conservation of energy demands that, when taken over any closed surface surrounding the scatterer, the integrals of the interference and scattered intensities must cancel:

$$\int_S dS_i F^s_i = -\int_S dS_i F^{int}_i. \tag{3.150}$$

Figure 26

The interference (left) in the forward direction only of scattered waves from incident compression waves and (right) the scattering in the forward direction and at the critical angle for incident shear waves.

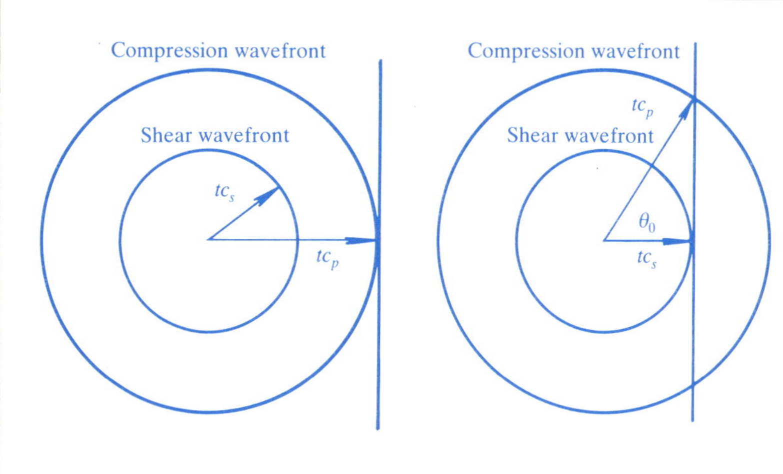

If the surface S is taken to be the limit of a large sphere as its radius tends to infinity, we have

$$\lim_{r\to\infty} r^2 \int d\Omega \hat{r}_i F_i^s = -\lim_{r\to\infty} r^2 \int d\Omega \hat{r}_i F_i^{int}, \tag{3.151}$$

in which the time average of the left-hand side is proportional to the total cross-section. The total cross-section is thus related to the interference of the incident power travelling parallel to the z axis with the scattered power travelling radially outwards from the flaw. The key difference between the acoustic and elastodynamic cases arises from the existence of two elastodynamic wave speeds. Figure 26 shows how this affects compression and shear wave scattering differently.

The explicit expression for the total scattering cross-section is

$$\sigma(\omega) = \frac{4\pi}{k_s} Im \frac{(\lambda+2\mu)k_s \tilde{P}_i(0)p_i^* + \mu k_s \tilde{S}_i(0)s_i^* + (\lambda+2\mu)k_p \tilde{P}_i(\theta_0)s_i^*}{k_p(\lambda+2\mu)p_i p_i^* + k_s \mu s_i s_i^*}. \tag{3.152}$$

In equation (3.152) the existence of additional interference at the critical angle

$$\theta_0 = \cos^{-1}(c_s/c_p), \tag{3.153}$$

when the incident wave is a shear wave, is quite clearly shown. The quantities denoted by a tilde are the azimuthal averages of the scattered amplitudes:

$$\tilde{\mathbf{A}}(\theta) = \frac{1}{2\pi}\int_0^{2\pi} \mathbf{A}(\theta,\phi)d\phi. \tag{3.154}$$

Note that the critical angle θ_0 defined in equation (3.152) is not the same as the critical angle found in reflection, as defined in equation (2.142).

In the discussion above it has been assumed that the waves are harmonic and that the properties of the scatterer are constant in time. Recently de Hoop (1985) has extended the work of Tan (1979) to derive theorems on the scattering and absorption of plane waves for arbitrary pulse shapes. Tan's theorems applied only to harmonic waves. At the same time, de Hoop considered time variations of the properties of the scatterer, and concluded that his theorems applied in such cases as well as in cases of non-linear behaviour in the scatterer.

3.2.2 The Kirchhoff approximation

As described above, the essence of the Kirchhoff approximation is that the values and gradients of the displacements are determined by the incident field, and that the incident field is non-zero only in regions which would be insonified in a geometric ray picture. The scattered wave is then calculated by means of a surface integral over the region of the scattering surface outside a sharply defined shadow zone. This assumption (Bouwkamp 1954) is not self-consistent: that is, the fields and gradients calculated at the surface using this assumption are not equal to those assumed. This is because it is not possible to specify simultaneously the stress and the displacement on a bounding surface; as shown in Chapter 2, either quantity on its own suffices.

In more recent applications (Beckmann and Spizzichino 1963; Trorey 1970) the basic assumption has been modified so that the outgoing wave from the scatterer is assumed to be the incident signal at each point multiplied by an appropriate plane-wave reflection coefficient. This assumption is the one now most commonly used. For elastic waves, the outgoing waves may be of compression and shear types, so that mode conversion can be included (Chapman 1984). This version of the Kirchhoff approximation, in which the integral on the scattering surface is taken over the geometrical elastodynamic fields, is called by some workers the physical elastodynamic approximation (Achenbach, Gautesen and McMaken 1982). If the scattered wave is taken from the integral in equation (3.134), then the boundary conditions on the surface can be incorporated. For a void, the stress at the surface is zero, and the displacement on the insonified surface is taken to be the sum of the displacement in the incident wave and the displacement in the reflected wave. This is the equivalent of the Rayleigh assumption of optics. In optics, the original Kirchhoff assumption gives a solution which is the average of the Rayleigh results for a perfectly soft and a perfectly hard boundary. The effect of the Rayleigh assumption is that the field near the scattering surface is given the value it would have had if the surface had been flat. Meecham (1956) has examined this assumption, and concludes that it will lead to good approximations if the radii of curvature of the scatterers are large compared with the wavelength and the slopes of the scattering surfaces relative to the wavefront are small.

For a crack-like scatterer, the scattered field is often expressed in terms of jumps in the wavefield across the crack. If the scatterer is taken to be a cylindrical object with vanishingly small height, the surface integral may be written as a single integral over the insonified surface, using the fact that the normal to one surface of the crack is the negative of the normal to the other surface. Then, if Δ denotes the difference between a field variable on the upper and lower surfaces, the scattered field may be written as

$$\begin{aligned} u_i^s = & \int_{S'} c_{jkpq} \Delta u_{p,q}(\mathbf{r}') G_{ij}(\mathbf{r}|\mathbf{r}') \hat{n}_k \, dS' \\ & - \frac{\partial}{\partial r_q} \int_{S'} c_{jkpq} \Delta u_j(\mathbf{r}') G_{ip}(\mathbf{r}|\mathbf{r}') \hat{n}_k \, dS', \end{aligned} \tag{3.155}$$

where now the surface S' is taken to be the insonified part of the crack. Again, it is possible to specify jumps in displacement, gradient or both. Problems formulated in this way are known as *saltus* problems (from the Latin for a jump), and may be solved exactly in Kirchhoff or Rayleigh forms. They may not, however, correspond to the correct physics of the scattering problem. The attraction of this formulation is that the difference in displacement across the crack is a quantity known as the *crack-opening displacement* which is used in fracture mechanics. This offers the possibility of using results for crack-opening displacements for static systems as first approximations for the scattering of low frequency waves. This is the basis of the quasistatic approximation described below.

Several applications to the scattering of ultrasound (for example Neubauer 1963; Adler and Lewis 1976; Johnson 1976; Haines and Langston 1980; Lam and Tsang 1985) have considered the acoustic analogue. This approach follows that of Freedman (1962), who treated the scattering of sound from a rigid body. Freedman assumed that the field of the transducer, treated as a point source at the origin, could be written as

$$\phi^i(\omega,t,\mathbf{r}) = \phi^0 e^{i\omega t} \frac{e^{-ikr}}{r}, \tag{3.156}$$

and that the strength of the Huygens source on an element $d\mathbf{S}$ of the rigid surface is $-\nabla\phi^i \cdot d\mathbf{S}$. In the far field of the transducer, where

$kr \gg 1$, the back-scattered signal from that element is

$$d\phi^r(\omega,t,\mathbf{r}) = ik\frac{\mathbf{r}\cdot d\mathbf{S}}{r^3}\phi^0 e^{i\omega t} e^{-2ikr}. \tag{3.157}$$

The total back-scattered signal is then

$$\phi^r(\omega,t) = ike^{i\omega t}\int_S e^{-2ikr}\frac{\mathbf{r}\cdot d\mathbf{S}}{r^3}. \tag{3.158}$$

The integral is over the area insonified by the incident wave. The term $\mathbf{r}\cdot d\mathbf{S}/r^3$ is the solid angle $d\Omega$ subtended by the scattering surface at the transducer. It is clear that if a large area of the scatterer lies at a constant range r from the transducer, the phase factor in the integral in equation (3.158) will be constant and there will be a large contribution to the scattered signal. This is, of course, just the specular reflection. In addition, Freedman (1962) shows how discontinuities in the solid angle as a function of range will also give large contributions, as will discontinuities in the range derivatives of the angle. This provides a clear link with the strong diffracted signals from edges and other discontinuities of the scatterer. This leads to the idea of diffracted signals from clearly identifiable regions of the scatterer, and thence to the use of ultrasonic spectroscopy for defect characterization (for example, Gericke 1963, 1966). Similar ideas may be used directly in the time domain (Mills, 1974).

More recently, the correct Kirchhoff theory for elastodynamics has been applied to scattering from cracks (Coffey and Chapman 1983; Chapman 1984; Chapman and Coffey 1984). They show that good results are obtained close to the specular reflection direction, for scatterers a few wavelengths in size. This is a useful observation, as defect detection systems are usually designed to operate as close as possible to the specular direction, so as to obtain maximum response (the same arrangement is often not optimal, however, for defect sizing systems). Their results for the back-scattered shear wave from a smooth strip-like crack of width $2a$, for $k_s a = 10$, at several angles of incidence, are compared with the exact solution (Keogh 1983) in Figure 27. Away from the specular direction, the Kirchhoff approximation to the crack-opening displacement is less accurate, as shown in Figure 27: this loss of accuracy away from the specular direction was also noted by Bowman, Senior and Uslenghi (1969) in electromagnetic scattering.

The elastodynamic version of the Kirchhoff approximation has recently been applied by Brind, Achenbach and Gubernatis (1984) to scattering by a cylindrical cavity. They first calculated the geometrical elastodynamic solution, by using plane-wave reflection coefficients for each element of the cylindrical surface, and then substituted the result in the surface integral. As shown in Figure 28, the simple geometric result is quite adequate for back scattering ($\theta > 90°$). For forward scattering the Kirchhoff approximation is an improvement over the

Figure 27

The back-scattered shear wave from a smooth strip-like crack of width $2a$, for $k_s a = 10$, as a function of angle of incidence. The exact solution is shown as a solid line, the Kirchhoff result as a dashed line. After Chapman (1984).

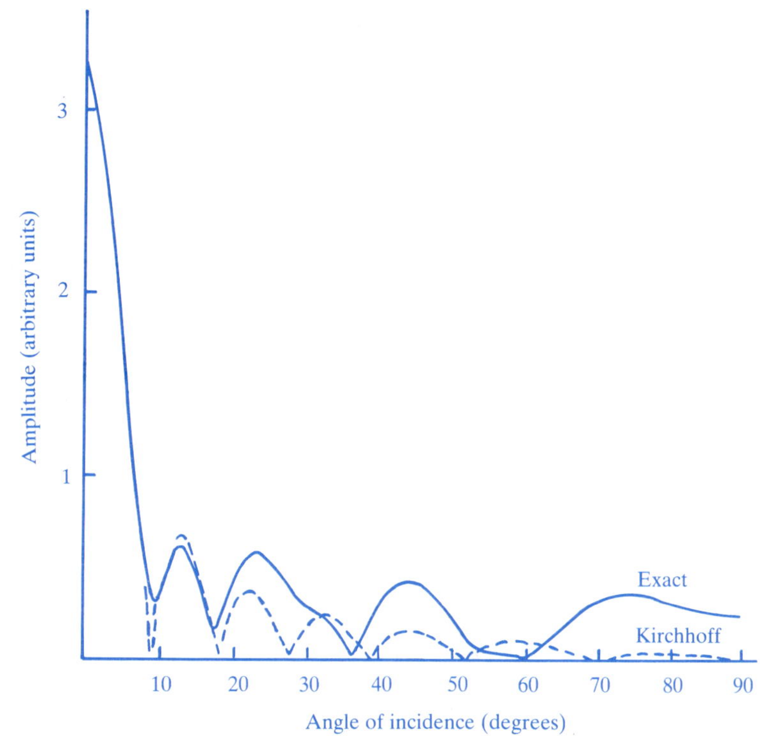

geometric approximation, but still does not agree well with the exact result. In fact, as shown by Brind, Achenbach and Gubernatis (1984), the discrepancy can largely be attributed to creeping waves on the surface of the cavity, which propagate the disturbance into the shadow zone. Because of the smoothness of the surface, these waves traverse the geometrical shadow boundary very efficiently. In a hybrid theory these rays are included in the surface integral, and good agreement with exact calculations is obtained in both forward and back scattering, although spurious oscillations are found near the shadow boundaries. Brind, Achenbach and Gubernatis (1984) describe a hybrid theory involving interpolation across the shadow boundary which gives excellent results. As this hybrid scheme requires about the same amount of computation as the t-matrix approaches (described below) for moderate sizes of cavity, the t-matrix methods are to be preferred except for high frequencies if forward scattering is important. If only back scattering is important, the geometrical or Kirchhoff approximations seem to be very reliable for smooth scatterers.

Figure 28

The compression (left) and shear (right) wave amplitude scattered when a compression wave is incident on a cylindrical cavity. The radius a of the cavity is such that $k_p a = 10$, and the figure shows exact (solid line), geometric (dotted) and Kirchhoff (dashed) results. After Brind, Achenbach and Gubernatis (1984).

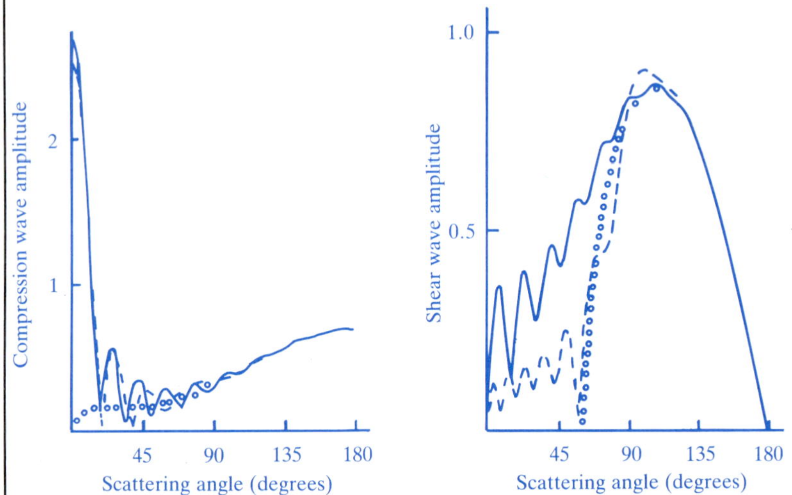

Although we have only discussed the single frequency Kirchhoff theory, the method can also be applied in the time domain. Early applications in acoustics (Friedman and Shaw 1962) used this approach, and Ko and Karlsson (1967) calculated the scattering of a compression wave from a rigid cylinder using an elastodynamic time domain Kirchhoff method. The direct time domain problem is particularly well suited to problems involving shock waves.

3.2.3 Weak scattering

If the scatterer has elastic properties which differ only slightly from those of the bulk material, an approximate solution of the integral equation (equation (3.129)) suggests itself. This relies on the fact that, if the scattering is weak, the total wavefield will be dominated by the incident wave. It is then possible to expand the scattered wave as a perturbation series (the Neumann or Born series). If we write the integral equation as

$$u_k(\mathbf{r}) = u_k^i(\mathbf{r}) - \varsigma \int_{V'} (\delta\rho\, \omega^2 G_{ki}(\mathbf{r}|\mathbf{r}') u_i(\mathbf{r}') - \delta c_{ijmn} u_{m,n}(\mathbf{r}') G_{ki,j}(\mathbf{r}|\mathbf{r}')) dV' \qquad (3.159)$$

where ς is a parameter which will be set equal to unity, and write the total displacement field as

$$u_k(\mathbf{r}) = \sum_{n=0}^{\infty} \varsigma^n u_k^{(n)}(\mathbf{r}), \tag{3.160}$$

then substituting (3.160) into (3.159) and equating powers of ς gives

$$U_k^{(0)}(\mathbf{r}) = u_k^i(\mathbf{r}). \tag{3.161}$$

The zero order field is therefore equal to the incident field. The higher powers of ς give

$$u_k^{(n)}(\mathbf{r}) = \int_{V'} (\delta\rho\, \omega^2 G_{ki}(\mathbf{r}|\mathbf{r}')u_i^{(n-1)}(\mathbf{r}') \\ -\delta c_{ijmn} u_{m,n}^{(n-1)}(\mathbf{r}')G_{ki,j}(\mathbf{r}|\mathbf{r}'))dV'. \tag{3.162}$$

In principle, equation (3.162) gives a scheme for an infinite sequence of iterations. In practice, the sequence is rarely taken beyond the first order term, so that the scattered wave is taken to be

$$u_k^s(\mathbf{r}) \\ = \int_{V'} (\delta\rho\, \omega^2 G_{ki}(\mathbf{r}|\mathbf{r}')u_i^i(\mathbf{r}') - \delta c_{ijmn} u_{m,n}^i(\mathbf{r}')G_{ki,j}(\mathbf{r}|\mathbf{r}'))dV', \tag{3.163}$$

that is, the integral is taken over the fields which would have occupied the volume of the scatterer if the scatterer had been absent. This is known as the first Born approximation, or the Rayleigh–Gans approximation.

For an incident plane wave with wavevector $\mathbf{k}^i$ and displacement vector $\mathbf{u}^i$, the scattering may be described, as above, in terms of the vector $\mathbf{f}(\mathbf{k}^s)$, giving

$$f_i(\mathbf{k}^s) = \frac{|k^s|^3}{4\pi\rho\omega^2}\left[\delta\rho\omega^2\delta_{il} - \delta c_{ijkl} k_j^s k_k^i\right] u_l^i S(\Delta\mathbf{k}). \tag{3.164}$$

Here the shape of the scatterer enters only through the shape factor S, given by

$$S(\Delta\mathbf{k}) = \int_{V'} e^{i\Delta\mathbf{k}\cdot\mathbf{r}}\, dV', \tag{3.165}$$

which is simply the Fourier transform of the shape.

In a simple two-dimensional example, if the scatterer is a cylinder of radius a the shape factor $S(\Delta\mathbf{k})$ is

$$S(\Delta\mathbf{k}) = \pi a^2 \frac{J_1(|\Delta\mathbf{k}|a)}{|\Delta\mathbf{k}|a}, \tag{3.166}$$

showing that the back-scattered cross-section, for which $|\Delta\mathbf{k}| = 2|\mathbf{k}|$, should exhibit oscillations corresponding to the oscillations of the first order Bessel function J_1. This should be contrasted with the result of the Kirchhoff approximation. For small scatterers, the shape factor reduces to a simple area

$$S(0) = \pi a^2 \tag{3.167}$$

and the cross-section varies as ω^4, in accordance with the Rayleigh theory. If we consider an isotropic solid, an incident compression wave will give rise to scattered compression and shear waves with the following variation with the angle θ between the incoming compression wave and the scattered wave:

$$\mathrm{P} \sim \left\{\cos\theta - \frac{\lambda+2\mu\cos^2\theta}{\lambda+2\mu}\right\} S(\Delta\mathbf{k}) \tag{3.168}$$

$$\mathrm{S} \sim \left\{\sin\theta - \sqrt{\frac{\mu}{\lambda+2\mu}}\sin 2\theta\right\} S(\Delta\mathbf{k}). \tag{3.169}$$

Models which are based on acoustic approximations, and which neglect the possibility of mode conversion, would give quite different angular variations: for example, they would replace the $\cos^2\theta$ in the compression wave scattering by unity (see, for example, Aks and Vezzetti 1980). This directional weighting difference between solids and fluids was stressed by Rose (1975).

The Born approximation is limited to small changes in material properties (as is to be expected from its character as the first term in a perturbation expansion). It is also limited to small scatterers for which $k^i a \leqslant 1$: this too is expected as only for small scatterers can it be accurate to replace the total field with the incident field. Detailed comparisons of the Born approximation with the exact result (Ying and Truell 1956; Einspruch, Witterholt and Truell 1960) have been made by Gubernatis, Domany, Krumhansl and Huberman (1977).

Figure 29 shows that the overall agreement between the Born approximation and the exact result for voids is rather poor. This is because the Born approximation cannot adequately model the forward scattering, which is strong in this case as the void is expected to create a shadow zone, where the scattered wave cancels the incident wave. For the mode-converted scattering, the agreement is better. It should be noted that the Born approximation does not conserve energy: the optical theorem (3.152) is not satisfied.

Gubernatis, Domany, Krumhansl and Huberman (1977) show that, away from the forward direction, the Born result is quite good. Figure 30 shows their results for sideways and back scattering. Further, they show that very good agreement with the exact result is obtained when the scattering is weaker, as for an inclusion of aluminium or steel in titanium. The frequency dependence, however, is not well represented for flaw dimensions larger than a fraction of a wavelength, as may be seen from the positions of the maxima and minima of scattering amplitude in Figure 30. It is clear from our earlier discussion of the

Figure 29

Amplitudes of scattered compression (a) and shear waves (b), for incident compression waves scattered from spherical voids in titanium. The figure shows the exact results and those calculated in the Born approximation, and σ_{ij} denotes the cross section for scattering from mode i into mode j (after Gubernatis, Domany, Krumhansl and Huberman (1977)).

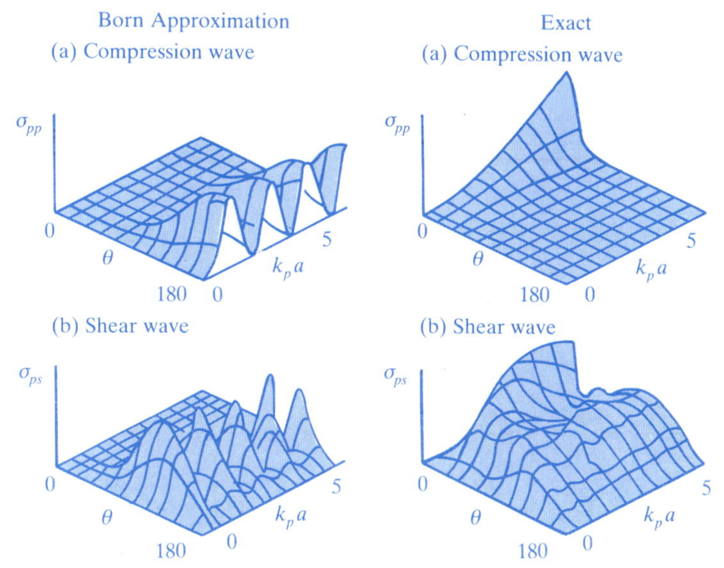

resonant behaviour of cavities: the frequency dependence of the scattering depends on the natural frequencies of the cavity, and those frequencies are not given by the zeros of the first order Bessel function. Thus the Born theory is not only a weak scattering approximation, it is also a long wavelength approximation.

3.2.4 Scattering from small defects

The theory of the scattering of ultrasound by defects which are small compared with the ultrasonic wavelength may be traced back to Rayleigh. His arguments were based on dimensional analysis, and although they provide insight into the frequency dependence of the scattering for optical and acoustic radiation they cannot predict the relative amounts of compression and shear waves that will be scattered in an elastic medium.

The theory was recently extended by Gubernatis (1979): the method described there involved the use of the quasistatic approximation, in that the variation of the sound field over the defect was assumed to be small, so that the shape of the response could be calculated by using formulae from static elasticity. In the quasistatic approximation, the

Figure 30

Amplitudes of scattered compression (a) and shear waves (b) at scattering angles greater than 90°, for incident compression waves scattered by spherical voids in titanium. The figure shows the exact results and those calculated in the Born approximation, and σ_{ij} denotes the cross section for scattering from mode i into mode j (after Gubernatis, Domany, Krumhansl and Huberman (1977)).

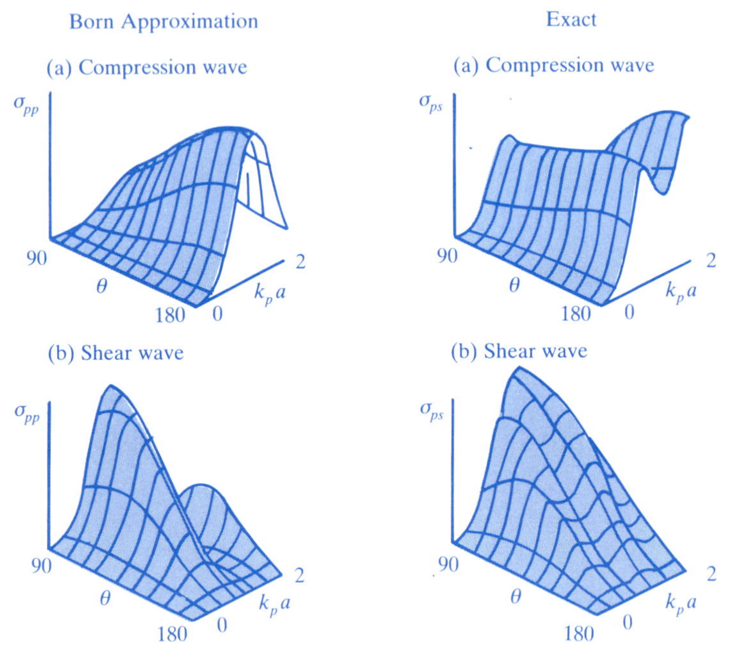

strain field inside the flaw is approximated by the static strain field, and in the extended quasistatic approximation the static field is multiplied by the phase factor $e^{i\mathbf{k}^i \cdot \mathbf{r}}$. Here we merely summarize the important equations, and refer the reader to the original literature for the details. The development of the theory may be found in Datta and Sangster (1974) and Datta (1977a, 1978), who used a method of matched asymptotic expansions, and Gubernatis (1979), who used perturbation theory. The final results apply only to the case of ellipsoidal flaws, but by making one axis infinitely long we may embrace cylindrical flaws, and by shrinking one axis to zero we may also include ribbon-like cracks or inclusions or flat elliptical defects.

Consider a flaw of volume V in an elastic solid: the solid has a density ρ and an elastic tensor c_{ijkl}, and the flaw has a density ρ' and elastic tensor c'_{ijkl}. We define the effect of the flaw in terms of the deviations

$$\delta\rho \equiv \rho' - \rho \qquad (3.170)$$

$$\delta c_{ijkl} \equiv c'_{ijkl} - c_{ijkl}. \qquad (3.171)$$

Then the scattered wave may be written in terms of the displacement vector $\mathbf{u}$ as

$$u_i^s \sim \frac{A_i}{k_p r} e^{ik_p r} + \frac{B_i}{k_s r} e^{ik_s r} \qquad (3.172)$$

in three dimensions and as

$$u_i^s \sim \frac{A_i}{\sqrt{(k_p r)}} e^{ik_p r} + \frac{B_i}{\sqrt{(k_s r)}} e^{ik_s r} \qquad (3.173)$$

in two dimensions, where the distance from the flaw to the point of observation is r and the compression and shear wavevectors are k_p and k_s. If the unit vector along the direction of observation of the flaw is $\hat{\mathbf{r}}$, then the scattered wavevectors for compression and shear waves may be written as $k_p\hat{\mathbf{r}}$ and $k_s\hat{\mathbf{r}}$, and $\mathbf{A}$ and $\mathbf{B}$ are defined by

$$A_i = \hat{r}_i \hat{r}_j f_j(k_p\hat{\mathbf{r}}) \qquad (3.174)$$

$$B_i = (\delta_{ij} - \hat{r}_i \hat{r}_j) f_j(k_s\hat{\mathbf{r}}), \qquad (3.175)$$

which serve to define the type of wave as compression or shear, and the scattering amplitude is

$$f_i(\mathbf{k}) = \frac{k^3}{4\pi\rho\omega^2}\left\{\omega^2 \delta\rho u_i^i + ik\hat{r}_j(\delta c_{ijkl}\Gamma_{klmn})\varepsilon_{mn}^i\right\} S(\Delta\mathbf{k}). \qquad (3.176)$$

Here $\mathbf{u}^i$ is the incident displacement, and the quantity Γ is related to the static Green tensor of the ellipsoid. The details are:

$$\Gamma_{ijkl} = (E - G\delta c)^{-1}_{ijkl} \qquad (3.177)$$

where E is the fourth rank identity tensor

$$E_{ijkl} = \tfrac{1}{2}(\delta_{ik}\delta_{jl} + \delta_{il}\delta_{jk}) \qquad (3.178)$$

and the only components of G which are required are

$$G_{1111} = Qa^2 I_{aa} + \mathrm{P} I_a \qquad (3.179)$$

$$G_{1122} = Qa^2 I_{ab} - \tfrac{1}{3} Q I_a \qquad (3.180)$$

$$G_{1212} = \tfrac{1}{2}(Q(a^2 + b^2) I_{ab} + R(I_a + I_b)) \qquad (3.181)$$

and terms related to these by symmetry. Here the principal axes of the ellipsoid are a, b and c, and the other terms are given by Eshelby (1957) and involve integrals over the shape of the defect. Γ represents the static response of the scatterer to the applied field.

The shape factor, which represents the dynamic part of the interaction, depends on the incident and scattered wavevectors $\mathbf{k}^i$ and $\mathbf{k}^s$ in the extended quasistatic approximation, as for the Born approximation, and if we take the incident wavevector to lie along the z axis it is given by

$$S(\mathbf{k}^s - \mathbf{k}^i) = 3V\frac{\sin x - x\cos x}{x^3} \tag{3.182}$$

$$x^2 = (k_1^s - k_1^i)^2 a^2 + (k_2^s - k_2^i)^2 b^2 \tag{3.183}$$

in three dimensions, and in two dimensions by

$$S(\mathbf{k}^s - \mathbf{k}^i) = A\frac{J_1(x)}{x}, \tag{3.184}$$

where V is the flaw volume, A its cross-sectional area. In the simple quasistatic approximation, the shape factors reduce to the volume and area respectively, with no dependence on wavevector.

Applications of quasistatic theory to the scattering by cracks have been numerous. The aims have been to extract parameters of relevance to fracture mechanics from the long wavelength scattering from cracks. Teitel (1978) and Domany, Teitel and Krumhansl (1978) examined the scattering by elliptical and circular cracks. Gubernatis, Krumhansl and Thomson (1979) and Budiansky and Rice (1978) made an interesting advance, by suggesting how the stress intensity factor (the key quantity in fracture mechanics) could be extracted directly from the scattering without performing a full analysis of the shape of the crack.

Figure 31

Amplitudes of scattered compression and shear waves, for incident compression waves, for a small spherical void, calculated in the quasistatic approximation: σ_{ij} is the cross section for scattering from mode i to mode j (after Gubernatis 1979).

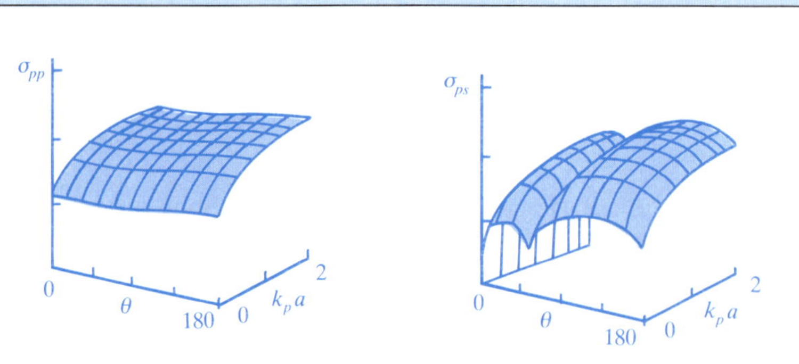

Figure 32

Amplitudes of scattered compression and shear waves, for incident shear waves, for a small spherical void, calculated in the quasistatic approximation: σ_{ij} is the cross section for scattering from mode i to mode j (after Gubernatis 1979).

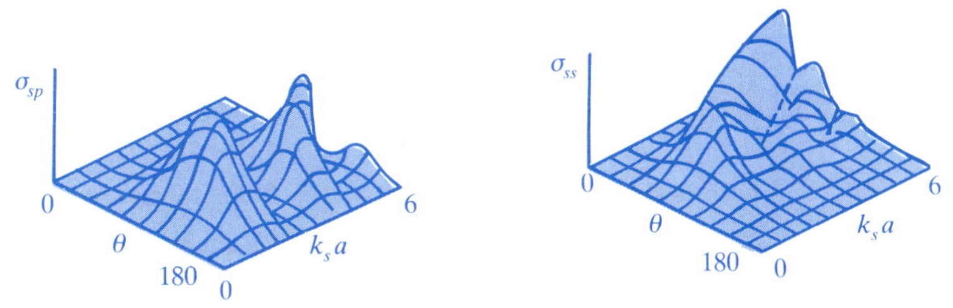

Figure 33

Amplitudes of scattered compression (solid line) and shear (dashed line) waves, for incident shear waves, for a small circular crack, calculated in the quasistatic approximation. The angles of incidence and scattering are both 15°. The units of amplitude are arbitrary, and the frequency axis is in units of c_s/a where a is the radius of the defect and c_s is the shear wave speed.

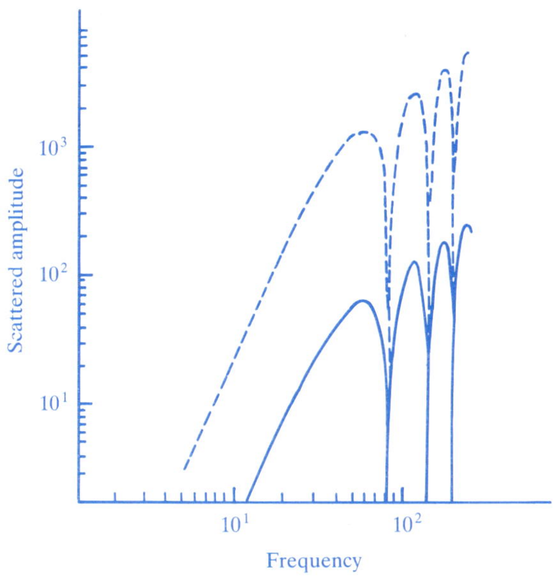

Examples of the angular distributions of scattered waves, and the frequency and size dependence of the scattering, are given in Figures 31

to 33. It is notable that the angular variation of the scattering in this limit is smooth, while the size dependence conforms to the Rayleigh scattering law for values of *ka* up to about $\frac{1}{3}$. It is important that the quasistatic approximation removes the deficiency of the Born approximation in the forward direction, and Gubernatis (1979) showed that the extended quasistatic approximation is in excellent agreement with the exact result for values of *ka* up to about unity.

Some simplifications may be made when scattering is confined to one plane which is a plane of symmetry of the defect, and of the form of the tensor Γ for small crack-like defects. Richardson (1984) has given a more general expansion of the scattering in powers of frequency, showing that the quasistatic approximation corresponds to the second degree scattering amplitude, and that the third degree amplitude vanishes for scatterers with inversion symmetry. The quasistatic approximation suitable only for defects with linear dimensions which are about one tenth of a wavelength or less. For larger crack-like defects the methods of the next section are appropriate.

3.2.5 Expansion methods and t-matrix methods

We have already described the use of expansion methods for problems in which the surface of the scatterer was a surface of constant value of one of the coordinates. In general the scatterer will not have such a convenient shape, but the idea of eigenfunction expansions is still useful, as the order to which the expansion is taken controls the amount of detail that can be included in the scattered field. In applications, as was the case with eigenfunction solutions which were exact, the method results in a set of matrix equations.

General scattering theory is often described in terms of channels and matrices. The terminology comes from elementary particle and high energy physics, where most of the information about the particles is obtained from scattering cross-sections. Well before and well after the scattering event, the particles are described by functions which resemble waves: the waves before and after the scattering are related by the scattering operator $\mathfrak{S}$. If the waves are expanded in, for example, plane waves, then the coefficients before and after the scattering are related by the matrix **S**. In general, the scattering is weak, and so the major part of $\mathfrak{S}$ describes the particles which pass through unscattered. It is then more useful to separate out the part of $\mathfrak{S}$ which corresponds to a change of state, and this contribution is the transition operator $\mathfrak{T}$, giving rise to the matrix **T**. The scattering and transition matrices are frequently referred to simply as the s-matrix and the t-matrix. Scattering theory is described in books by Newton (1966) and Taylor (1972).

The t-matrix theory of scattering may be derived from Green functions. The technique relies on the expansion of the Green function in basis functions. We shall show first how the method applies to scalar waves, and then derive the corresponding expressions for elastic waves. Consider first the two-dimensional case. It is well known (see, for example, Morse and Feshbach 1953, page 1372) that the Green function in free space may be expanded as

$$g(\mathbf{r}|\mathbf{r}') = i\pi H_0^{(1)}(kR) i\pi \sum_{m=0}^{\infty} \cos m\theta \, J_m(kr_<) H_m^{(1)}(kr_>) \tag{3.185}$$

where we have introduced $r_>$, the greater of r and r', and $r_<$, the lesser of the two. θ is the angle between $\mathbf{r}$ and $\mathbf{r}'$. In terms of the individual azimuthal angles of $\mathbf{r}$ and $\mathbf{r}'$,

$$\begin{aligned} g(\mathbf{r}|\mathbf{r}') = i\pi\Big\{ & \sum_{m=0}^{\infty} \cos m\theta_< J_m(kr_<) \cos m\theta_> H_m^{(1)}(kr_>) \\ & + \sum_{m=1}^{\infty} \sin m\theta_< J_m(kr_<) \sin m\theta_> H_m^{(1)}(kr_>)\Big\}. \end{aligned} \tag{3.186}$$

We may write this as

$$g(\mathbf{r}|\mathbf{r}') = i\pi \sum_\sigma \psi_\sigma(kr_>) \mathcal{R}[\psi_\sigma(kr_<)] \tag{3.187}$$

where σ is an index which refers to both the order m and the symmetry sin or cos, and $\mathcal{R}$ denotes the real part. Note that ψ_σ will be singular at the origin, because of the singularity in $H_m^{(1)}$, but $\mathcal{R}[\psi_\sigma(kr_<)]$ will be regular there.

The t-matrix method depends on a suitable choice of expansion functions ψ_σ. The incident field is expanded in the regular functions $\mathcal{R}[\psi_\sigma(kr_<)]$ with coefficients a_σ, and the scattered field is expanded in the functions ψ_σ with coefficients f_σ. Then, by substituting these expansions in the integral equation and matching the two solutions (inside and outside the scatterer) on the boundary of the scatterer, the matrix $\mathbf{T}$ is the relationship between $\mathbf{f}$ and $\mathbf{a}$:

$$f_\tau = \sum_\sigma T_{\tau\sigma} a_\sigma. \tag{3.188}$$

$\mathbf{T}$ is defined in terms of surface integrals $\mathbf{Q}$ as

$$\sum_\sigma T_{\tau\sigma} Q_{\sigma\eta} = \mathcal{R}[Q_{\tau\eta}]. \tag{3.189}$$

For elastic waves, t-matrix theory was derived by Waterman (1976) and Varatharajulu and Pao (1976). They define vector basis functions $\mathbf{L}$ and $\mathbf{M}$ for the compression and shear waves. If we work again in two dimensions, in planes perpendicular to the z axis, and $\hat{\mathbf{r}}_3$ is a unit vector along the z axis,

$$\mathbf{L}_\sigma(\mathbf{r}) = \nabla \psi_\sigma(k_p \mathbf{r}) \tag{3.190}$$

$$\mathbf{M}_\sigma(\mathbf{r}) = \nabla_\wedge [\psi_\sigma(k_s \mathbf{r}) \hat{\mathbf{r}}_3] \tag{3.191}$$

where ψ_σ is defined above. The $\mathbf{L}$ and $\mathbf{M}$ functions have orthogonality relations, which are tabulated by Varatharajulu and Pao. The incident and scattered waves are now expanded as

$$\mathbf{u}^i(\mathbf{r}) = \sum_\sigma \{a_\sigma \mathcal{R}[\mathbf{L}_\sigma(\mathbf{r})] + b_\sigma \mathcal{R}[\mathbf{M}_\sigma(\mathbf{r})]\} \tag{3.192}$$

and

$$\mathbf{u}^s(\mathbf{r}) = \sum_\sigma \{f_\sigma \mathbf{L}_\sigma(\mathbf{r}) + g_\sigma \mathbf{M}_\sigma(\mathbf{r})\}. \tag{3.193}$$

Here the expansion coefficients a and b are known, f and g are to be found. The Green tensor can also be expanded in terms of the **L** and **M** functions, giving

$$\begin{aligned} G_{ij}(\mathbf{r}|\mathbf{r}',\omega) = \frac{\pi}{\rho\omega^2}\sum_\sigma \{ & L_{\sigma i}(\mathbf{r}_>)\mathcal{R}[L_{\sigma j}(\mathbf{r}_<)] \\ & + M_{\sigma i}(\mathbf{r}_>)\mathcal{R}[M_{\sigma j}(\mathbf{r}_<)]\}. \end{aligned} \tag{3.194}$$

Note that G_{ij} has now been separated into longitudinal and transverse parts, with appropriate wavevectors k_p and k_s (Morse and Feshbach 1953). The stress Green tensor may be expanded in a similar way.

Now consider a region inside the scatterer, where $|\mathbf{r}|<|\mathbf{r}'|$. The integral equation gives

$$\begin{aligned} & \sum_\sigma \{a_\sigma \mathcal{R}[L_{\sigma i}(\mathbf{r})] + b_\sigma \mathcal{R}[M_{\sigma i}(\mathbf{r})]\} \\ & = -\sum_\sigma \sum_{j,k} \int_S \{\Sigma_{ijk}(\mathbf{r}|\mathbf{r}')u_j(\mathbf{r}') - G_{ij}(\mathbf{r}|\mathbf{r}')\tau_{jk}(\mathbf{r}')\} dS'_k. \end{aligned} \tag{3.195}$$

Outside the scatterer we have instead:

$$\begin{aligned} & \sum_\sigma \{f_\sigma L_{\sigma i}(\mathbf{r}) + g_\sigma M_{\sigma i}(\mathbf{r})\} \\ & = -\sum_\sigma \sum_{j,k} \int_S \{\Sigma_{ijk}(\mathbf{r}'|\mathbf{r})u_j(\mathbf{r}') - G_{ij}(\mathbf{r}'|\mathbf{r})\tau_{jk}(\mathbf{r}')\} dS'_k. \end{aligned} \tag{3.196}$$

The change from $(\mathbf{r}|\mathbf{r}')$ to $(\mathbf{r}'|\mathbf{r})$ between the two last equations indicates that, whereas in the first equation $\mathbf{r}$ was the argument of $\mathcal{R}[\mathbf{L}]$ and $\mathcal{R}[\mathbf{M}]$, in the second equation $\mathbf{r}$ is the argument of **L** and **M**. In both equations the displacement $\mathbf{u}_j$ and surface tractions $\tau_{jk} dS_k$ at the surface remain unspecified. If we take a scatterer which is an empty cavity, the tractions will be zero. The displacements on the surface of the cavity will still be unknown, so we expand them in terms of the basis functions as

$$\mathbf{u}(\mathbf{r}') = \sum_\sigma \{\alpha_\sigma \mathcal{R}[\mathbf{L}_\sigma(\mathbf{r})] + \beta_\sigma \mathcal{R}[\mathbf{M}_\sigma(\mathbf{r})]\}. \tag{3.197}$$

We can now reduce our equations to matrix form, by taking scalar products of the expansions of the integral equations with the basis functions $\mathcal{R}[\mathbf{L}]$, $\mathcal{R}[\mathbf{M}]$ or **L**, **M**, as appropriate. Using the orthogonality relations, we find that

$$\begin{Bmatrix} a_\sigma \\ b_\sigma \end{Bmatrix} = -i\begin{Bmatrix} Q_{\sigma\tau}^{LL} & Q_{\sigma\tau}^{LM} \\ Q_{\sigma\tau}^{ML} & Q_{\sigma\tau}^{MM} \end{Bmatrix}\begin{Bmatrix} \alpha_\tau \\ \beta_\tau \end{Bmatrix} \tag{3.198}$$

and, for the expansion external to the scatterer,

$$\begin{Bmatrix} f_\sigma \\ g_\sigma \end{Bmatrix} = i\mathcal{R}\begin{Bmatrix} Q_{\sigma\tau}^{LL} & Q_{\sigma\tau}^{LM} \\ Q_{\sigma\tau}^{ML} & Q_{\sigma\tau}^{MM} \end{Bmatrix}\begin{Bmatrix} \alpha_\tau \\ \beta_\tau \end{Bmatrix}. \tag{3.199}$$

The **Q** matrices now contain all the information about the scattering. The elements of the **Q** matrices are integrals of products of the basis functions over the surface. For the cavity, the explicit forms are

$$Q_{\sigma\tau}^{LL} = \frac{\pi}{\rho\omega^2}\int_S \sum_{ij} \mathcal{R}[L_{\tau i}(\mathbf{r}')]\Big\{\lambda\delta_{ij}\sum_k \frac{\partial}{\partial r_k} L_{\sigma k}(\mathbf{r}') + \mu\frac{\partial}{\partial r'_i}L_{\sigma j}(\mathbf{r}') + \mu\frac{\partial}{\partial r'_j}L_{\sigma i}(\mathbf{r}')\Big\} dS'_j \tag{3.200}$$

and

$$Q_{\sigma\tau}^{ML} = \frac{\pi\mu}{\rho\omega^2}\int_S \sum_{ij} \mathcal{R}[L_{\tau i}(\mathbf{r}')]\Big\{\frac{\partial}{\partial r'_i}M_{\sigma j}(\mathbf{r}') + \frac{\partial}{\partial r'_j}M_{\sigma i}(\mathbf{r}')\Big\} dS'_j. \tag{3.201}$$

$\mathbf{Q}^{LM}$ is obtained by replacing $\mathcal{R}[\mathbf{L}]$ by $\mathcal{R}[\mathbf{M}]$ in the first of these equations, and $\mathbf{Q}^{MM}$ by replacing $\mathcal{R}[\mathbf{L}]$ by $\mathcal{R}[\mathbf{M}]$ in the second. Then a simple elimination of $[\alpha,\beta]$ yields the solution we gave before:

$$\sum_\sigma T_{\tau\sigma} Q_{\sigma\eta} = \mathcal{R}[Q_{\tau\eta}]. \tag{3.202}$$

In three dimensions the principles of the method are exactly the same. Three sets of basis functions are required, to cater for the compression waves and two polarizations of shear waves, and the angular variations are described by spherical harmonics $Y_{lm}(\theta,\phi)$. This leads to a rather more compact notation than the real spherical harmonics used by Varatharajulu and Pao (1976). The basis functions may be taken as

$$L_\sigma(\mathbf{r}) = \nabla h_n^{(1)}(k_p r) Y_\sigma(\theta,\phi) \tag{3.203}$$

$$M_\sigma(\mathbf{r}) = \frac{1}{\sqrt{(l(l+1))}}\nabla_\wedge\left[\mathbf{r} h_n^{(1)}(k_s r) Y_\sigma(\theta,\phi)\right] \tag{3.204}$$

$$N_\sigma(\mathbf{r}) = \frac{1}{\sqrt{(l(l+1))}}\nabla_\wedge\nabla_\wedge\left[\mathbf{r} h_n^{(1)}(k_s r) Y_\sigma(\theta,\phi)\right], \tag{3.205}$$

where the subscript σ stands for the pair (l,m).

In problems which include several different scattering centres it is necessary to consider the multiple scattering among the different centres. This is more complicated because, in the expansion of the displacements, it is necessary to know how to translate the basis functions; that is, it is necessary to know how to expand a basis function centred on one scatterer about the centre of another scatterer. The resulting matrix equations have been studied by Peterson and Strom (1973, 1974a), Bostrom (1980) and Varadan and Varadan (1981). Domany, Entin–Wohlman and Mizrachi (1984) have also considered multiple scattering of elastic waves, and find that their expansion in terms of single-scatterer t-matrices may be truncated at the double-scattering term. From this they conclude that the interaction terms will generally be small enough for the method to be applicable to a scatterer near a surface (Domany and Entin–Wohlman 1984). For SH wave scattering, defects at surfaces can be treated by using image theory (Varadan, Wall, Tsao and Varadan 1982), but a similar trick is not possible for P and SV scattering. Varadan, Tsao and Varadan (1984) have applied a multiple-scattering t-matrix approach to subsurface inclined cracks, and show that the multiple scattering becomes important when the depth of the crack below the surface is less than its diameter. Peterson and Strom (1974b, 1975) have also applied the matrix method to layered systems.

3.2.6 Practical aspects of the t-matrix method

The expressions derived in the previous section are quite general, and in particular the sum over the index σ is taken over a full set of expansion functions. In other words, we have only obtained a formal solution which still contains an infinite summation. The convergence of the expansion, in a formal sense, has been confirmed by various authors: clear expositions are given by Millar (1973), Bates and Wall (1977), Waterman (1969) and Pao (1978).

In any real calculation we have to use a finite set of functions **L**, **M**, and in general the integrals involved in **Q** will have to be done numerically. The numerical integrations in themselves pose no difficulty. As Waterman (1969) pointed out, one advantage of the t-matrix formulation is that the surface integrals needed involve non-singular kernels. This is in contrast with variational methods, where multiple surface integrals with singular kernels have to be evaluated. General approaches to the construction of t-matrix theories have been given by Visscher (1980a,b). The method of Waterman (1969) for acoustic scattering was extended to elastic waves by Waterman (1976) and Varatharajulu and Pao (1976).

The simplest way of obtaining a finite, soluble, matrix equation is simply to truncate the series at some finite index σ. The accuracy of this simple truncation method may be assessed by comparing the results obtained with those arising from the solution with an extra term in the series. Waterman (1969, 1971, 1976) has proposed a modification of the numerical method which can be applied if the truncated matrix representation of the scattering operator $\mathfrak{S}$ satisfies the requirements of time reversal and energy conservation. This requires the matrix **S** to be symmetric and unitary:

$$S^T = S \qquad \text{and} \qquad S^{-1} = S^*. \tag{3.206}$$

It is not immediately obvious that forcing the truncated matrix to obey these physical constraints will enhance the convergence of the method, though Waterman (1969, 1971) finds that if the symmetry requirement is met then improved convergence is obtained for some electromagnetic problems. In his computations, however, he tested the matrix for symmetry and took advantage of the symmetry if it was present, rather than imposing the symmetry as an additional constraint.

Visscher (1980a,b) has developed the Method Of Optimal Truncation (MOOT), in which he recognizes that truncation must lead to an inexact satisfaction of the boundary conditions. He argues that the optimal truncation is the one which minimizes the difference between the required and actual conditions on the boundary of the scatterer. If the operator $\mathfrak{B}$ represents the boundary condition (for a cavity, $\mathfrak{B}$ will be the stress operator), the error in satisfying the boundary conditions may be quantified by

$$\Delta = \int_S |\mathfrak{B}\cdot\mathbf{u}|^2 dS, \tag{3.207}$$

with **u** being the sum of the expansions of the incident and scattered fields. The coefficients of the scattered field f_σ are then chosen so as to minimize Δ. The resulting matrix equations have the same form as those in Waterman's theory but differ in detail. They have the advantage of being derived from a minimum principle, and, as demonstrated by Visscher (1980a), of converging uniformly with increasing σ. A later development (Opsal and Visscher 1985) addresses the question of the existence of the solution in terms of a single expansion. For shapes which are far removed from the 'natural' shapes for the basis functions of the expansion, they advocate the use of different expansions in different regions. As with the simple truncation method, time reversal symmetry and energy conservation are not guaranteed in the method of optimal truncation, but Visscher specifically advises against modifying the theory to include them. There is still controversy on this point (Waterman 1980).

It is possible to estimate the order of magnitude of the computational effort required to solve a scattering problem by the t-matrix method. If we consider the scattering by a cylinder of radius a, then to describe creeping waves on the surface one requires angular variations of the form $\cos l_{\max}\theta$ and $\sin l_{\max}\theta$, with $l_{\max} \approx ka$. This leads to square matrices which are $(2l_{\max}+1)\times(2l_{\max}+1)$ in two dimensions. In three dimensions each l value has $2l+1$ values of m associated with it, and there are three different potential functions, so the matrices are $3(l_{\max}+1)^2\times 3(l_{\max}+1)$. Evidently this becomes excessive for large ka.

This estimate of the value of l_{max} becomes unreliable for scatterers with regions of large and small curvature (for example, ellipses or ellipsoids of large eccentricity, such as crack-like defects): this is described by Waterman (1969, 1971) and Bolomey and Wirgin (1974). The choice of basis functions is important: for example, if a spheroidal basis set is used, even small deviations from a spheroidal geometry lead to significant deteriorations in the numerical accuracy of the method. This point has recently been investigated by Hackman and Todroff (1985). For high aspect ratios, it seems (Tobocman 1985) that a method based directly on the Helmholtz integral equation may converge more rapidly than the t-matrix method.

A good guide to the accuracy of the optimal truncation method seems to be hard to find. Generally the approach used is to increase the order l_{max} and study the convergence of the method. One pointer to possible difficulties may be found in Visscher (1981), who presents results for a penny-shaped crack of radius a. For long wavelengths the computed results lie about 10 per cent higher than the Rayleigh result for $l_{max} = 10$, and the error is reduced to 8 per cent for $l_{max} = 14$. The problem arises from the singular nature of the stress field at the edge of the crack, which cannot be adequately modelled by the expansion functions. This means that the optimal truncation method should not be used for scatterers which are too small: but we have already discussed suitable methods for such cases.

As currently formulated, the t-matrix method is directly applicable only to problems with a periodic time variation. To treat pulses, the problem must be solved for many frequencies and the pulse synthesized by Fourier methods. Waterman (1980) has suggested that it might be possible to use a complete set of time-dependent basis functions for the expansion, and to solve general time-varying problems in that way, but this does not seem to have been attempted yet.

3.3 Geometric ray theory

In this section we shall try to relate the fundamental ideas of geometric acoustics to the detailed wave theory we have been using so far. Geometric acoustics is the acoustic analogue of geometric optics, which is encountered in every basic physics course. The problem is to relate the simple laws of reflection and refraction, which were derived for infinite plane wavefronts, to the more complicated wavefronts encountered in practice. Geometrical acoustics makes the generalization by treating *every point* on a wavefront as part of a plane wave. It assumes that rays and wavefronts are always perpendicular to each other, even in complicated cases where the material properties vary from point to point, so that the rays are not straight. Geometric acoustics is valid in the high-frequency limit: we shall see, in fact, that it represents the first term in an expansion in inverse powers of frequency.

The formalism of ray theory can be traced back to Hamilton's work in the early 19th century (see Whitham 1974, chapter 3 or Lighthill 1978, chapter 4). The disturbance, be it light or sound, is assumed to travel along curves called rays, and at an interface between two media rays are bent according to Snell's law or its acoustic analogue, giving reflected and refracted rays. Along each ray the phase, polarization and amplitude are taken from plane-wave theory and the conservation of energy. The theory was placed on a firmer footing for electromagnetic waves by Luneberg in 1944 (see Luneberg 1964; Kline and Kay 1965), and a clear derivation for acoustic waves is given by Friedlander (1958). Its more recent applications may be said to stem from the work of Keller (1957). As explained by Keller (1979), ray theory has similarities with Rayleigh–Gans theory, in which scatterers introduce phase shifts on rays. When applied to quantum mechanical scattering, with similar assumptions, the theory is known as the eikonal approximation: one of its most common forms is Glauber theory (Glauber 1959). Ray theory is also related to the asymptotic evaluation of the Feynman path integral which is used in quantum mechanical calculations (Morette 1951). Keller (1953) made an important extension to geometric ray theory by introducing rays which originate at edges of scattering bodies, and propagate into shadow zones: these he called *diffracted rays*. He justified the approach on physical grounds by describing it as an extension of Fermat's principle to diffraction.

The geometric theory of diffraction for elastic waves was derived by Karal and Keller (1959). Starting from the wave equation for an isotropic medium

$$(\lambda+\mu)u_{j,ji}+\mu u_{i,jj}+\rho f_i = \rho u_{i,tt}, \tag{2.39}$$

a time harmonic solution is sought, and is assumed to have the form

$$\mathbf{u} = \mathbf{A}e^{i\omega(S-t)}, \tag{3.208}$$

where $\mathbf{A}$ and S are both functions of position, to be determined. It is further assumed that S is independent of the frequency ω and that $\mathbf{A}$ may be expanded in the form

$$\mathbf{A} = \sum_{n=0}^{\infty}(i\omega)^{-n}\mathbf{A}^{(n)}. \tag{3.209}$$

Again the vectors $\mathbf{A}^{(n)}$ are functions of position to be determined. One may note in passing that (3.209) contains only integer powers of ω, whereas the exact theory of diffraction typically introduces half-integer powers of ω (as shown in the next section). This makes it clear that ray theory in this form cannot treat diffraction effects. On substituting equation (3.209) into (3.208), and the result into (2.39), one obtains a series of equations in various powers of ω. If equation (2.39) is to be satisfied for all values of ω, then it is possible to solve the equations for each power of ω separately. This gives, for each $n \geqslant 0$,

$$\begin{aligned}
&-\rho A_i^{(n)} + (\lambda+\mu)(A_j^{(n)} \nabla_j S)\nabla_i S + \mu(\nabla_j S)(\nabla_j S) A_i^{(n)} \\
&+ (\lambda+\mu)\nabla_i (A_j^{(n-1)} \nabla_j S) + (\lambda+\mu)(\nabla_j A_j^{(n-1)})\nabla_i S + 2\mu(\nabla_j S)\nabla_j A_i^{(n-1)} \\
&+ \mu(\nabla_j \nabla_j S) A_i^{(n-1)} + (\lambda+\mu)\nabla_i(\nabla_j A_j^{(n-2)}) + \mu \nabla_j \nabla_j A_i^{(n-2)} = 0.
\end{aligned} \tag{3.210}$$

The original expansion was in descending powers of ω, and the vectors $\mathbf{A}^{(n-1)}$ and $\mathbf{A}^{(n-2)}$ are both zero.

The first term in the series, $n=0$, may be simplified by taking the scalar product of equation (3.210) with the gradient of S, giving

$$[-\rho + (\lambda+2\mu)(\boldsymbol{\nabla} S)^2]\mathbf{A}^{(0)} \cdot \boldsymbol{\nabla} S = 0, \tag{3.211}$$

or the vector cross product with the gradient of S,

$$[-\rho + \mu(\boldsymbol{\nabla} S)^2]\mathbf{A}^{(0)}_{\wedge} \boldsymbol{\nabla} S = 0. \tag{3.212}$$

These two equations have two possible solutions. Either

$$\mathbf{A}^{(0)} \cdot \boldsymbol{\nabla} S = 0 \qquad \text{and} \qquad (\boldsymbol{\nabla} S)^2 = \rho/\mu \tag{3.213}$$

or

$$\mathbf{A}^{(0)}_{\wedge} \boldsymbol{\nabla} S = 0 \qquad \text{and} \qquad (\boldsymbol{\nabla} S)^2 = \rho/(\lambda+2\mu). \tag{3.214}$$

The equations for $\boldsymbol{\nabla} S$ are eikonal equations, which determine the phase S in terms of the phase on an initial surface. The initial surface may, of course, be an initial wavefront. The perpendiculars to the surfaces of constant phase are the rays, and from equations (3.212) and (3.214) the rays are straight lines. It is clear that equations (3.212) and (3.214) describe shear and compression waves respectively. It also follows that the phase factor S is a linear function of distance along a ray, s: for shear waves

$$\begin{aligned}
S &= S_0 + \sqrt{\frac{\rho}{\mu}}(s-s_0) \\
&= S_0 + (s-s_0)/c_s,
\end{aligned} \tag{3.215}$$

and for compression waves

$$\begin{aligned}
S &= S_0 + \sqrt{\frac{\rho}{\lambda+2\mu}}(s-s_0) \\
&= S_0 + (s-s_0)/c_p \cdot
\end{aligned} \tag{3.216}$$

It is now necessary to solve for the amplitudes $\mathbf{A}^{(n)}$. We consider only the first term, and refer the reader to Karal and Keller (1959) for higher orders. For shear waves, substituting equation (3.212) into equation (3.210) for $n=1$ gives

$$(\lambda+\mu)(\mathbf{A}^{(1)} \cdot \boldsymbol{\nabla} S)\boldsymbol{\nabla} S + (\lambda+\mu)(\boldsymbol{\nabla} \cdot \mathbf{A}^{(0)})\boldsymbol{\nabla} S + 2\mu(\boldsymbol{\nabla} S \cdot \boldsymbol{\nabla})\mathbf{A}^{(0)} + \mu \mathbf{A}^{(0)} \nabla^2 S = 0. \tag{3.217}$$

The scalar product of this equation with the gradient of S yields

$$\mathbf{A}^{(1)}\cdot\boldsymbol{\nabla}S = -\boldsymbol{\nabla}\cdot\mathbf{A}^{(0)} \tag{3.218}$$

which may be substituted back to give

$$2(\boldsymbol{\nabla}S\cdot\boldsymbol{\nabla})\mathbf{A}^{(0)}+(\nabla^2 S)\mathbf{A}^{(0)} = 0, \tag{3.219}$$

a first order linear partial differential equation for $\mathbf{A}^{(0)}$.

If we work in terms of distances along rays, and note that the expression $\boldsymbol{\nabla}S\cdot\boldsymbol{\nabla}$ is a derivative taken along a ray, we may write equation (3.219) as

$$\frac{d}{ds}\mathbf{A}^{(0)} = c_s(\nabla^2 S)\mathbf{A}^{(0)}, \tag{3.220}$$

which has the solution

$$\mathbf{A}^{(0)}(s) = \mathbf{A}^{(0)}(s_0)e^{-\frac{1}{2}c_s\int_{s_o}^{s}\nabla^2 S\,ds}. \tag{3.221}$$

Now Luneburg (1964) has shown that the integral in equation (3.221) may be written as

$$\int_{s_o}^{s}\nabla^2 S\,ds = \left[\frac{d\sigma(s_0)}{d\sigma(s)}\right]^{\frac{1}{2}}, \tag{3.222}$$

where σ is the cross-sectional area of a bundle of rays. As Karal and Keller (1959) point out, $d\sigma(s_0)/d\sigma(s)$ is the limit of the ratios of the areas at s_0 and s as the bundle of rays shrinks to a single ray. It is also the Jacobian of the mapping by rays of the area at s onto the area at s_0.

For practical purposes it is more convenient to express the area ratio in terms of local radii of curvature of the wavefront. If ρ_1 and ρ_2 are the principal radii of curvature at s_0 then the radii at s will be (ρ_1+s-s_0) and (ρ_2+s-s_0), so that

$$\mathbf{A}^{(0)}(s) = \mathbf{A}^{(0)}(s_0)\left[\frac{\rho_1\rho_2}{(\rho_1+s-s_0)(\rho_2+s-s_0)}\right]^{\frac{1}{2}}. \tag{3.223}$$

This shows that the wave amplitude varies as the square root of the cross-sectional area of a bundle of rays – exactly as is to be expected if the energy is assumed to propagate along the bundle. An analogous result is obtained for compression waves. The method may be extended to inhomogeneous media (Karal and Keller 1959), when it provides useful insights.

The application of geometric ray theory relies on the expansion of the initial wavefield into rays: this is a simple decomposition of the field. Each ray is then treated as though it formed part of an infinite plane wave of the appropriate amplitude. Thus the laws of plane-wave reflection and refraction are applied to each ray. When scatterers are present, their effects are again included by solving the canonical problem of a plane wave interacting with the scatterers, and incorporating the amplitude and phase from ray theory.

The diffraction from a semi-infinite crack is an important canonical problem. The full theory (discussed below) may be cast in the form of a geometric acoustics field (which is zero in the shadow region) and a diffracted field. The diffracted field is a cylindrical wave whose axis coincides with the tip of the crack. The diffracted field may be written in terms of a diffraction coefficient D, which gives the angular distribution, and the cylindrical wave:

$$\mathbf{u} = \mathbf{u}^{(0)} D(\theta,\phi) \frac{e^{-i\omega s/c_s}}{s^{\frac{1}{2}}}. \tag{3.224}$$

The further development of the cylindrical wave is then followed using ray theory, and the diffraction coefficient is a simple modification of the amplitude and phase.

The advantage of expressing diffraction in this way is that we can solve any diffraction problem, with an arbitrary number of edges, in the following way. We consider the geometric acoustics field, and add to it the field produced by any diffracting edges. Each diffracting edge is treated separately, and the expression for D inserted with appropriate weight. The essence of this approach is that diffraction is largely a *local* phenomenon. If an incident wave sees a diffracting edge which looks locally like the edge of an infinite plane, the solution for an infinite plane will be an accurate representation of the diffraction.

3.3.1 A canonical problem: scattering from crack edges

The theory of the reflection and diffraction of ultrasound at oblique incidence on crack-like discontinuities has been reviewed by Achenbach, Gautesen and McMaken (1982). The approach commonly involves the conversion of the dynamic elasticity problem into a form amenable to a solution using the Wiener–Hopf method, and leaves one relatively simple one-dimensional integral to be performed numerically. For completeness, the relevant equations are repeated below. We follow the method and notation of Maue (1953), but similar results have been obtained by Chapman (1981), Ogilvy and Temple (1983), and Achenbach and co-workers (see Achenbach, Gautesen and McMaken 1982).

Consider a semi-infinite crack in an infinite medium, where the crack is taken to have stress-free surfaces in the half-plane $y=0$, $x<0$, $-\infty \leqslant z \leqslant \infty$, as shown in Figure 34. Compression or shear waves with wavevectors in the (x,y) plane are incident on the crack. The angular frequency of the incident waves is taken to be ω. As well as the compression and shear wave speeds in the material c_p and c_s we use

Figure 34

The directions of the incident wave ϕ_0 and the diffracted wave θ for the scattering by a semi-infinite crack.

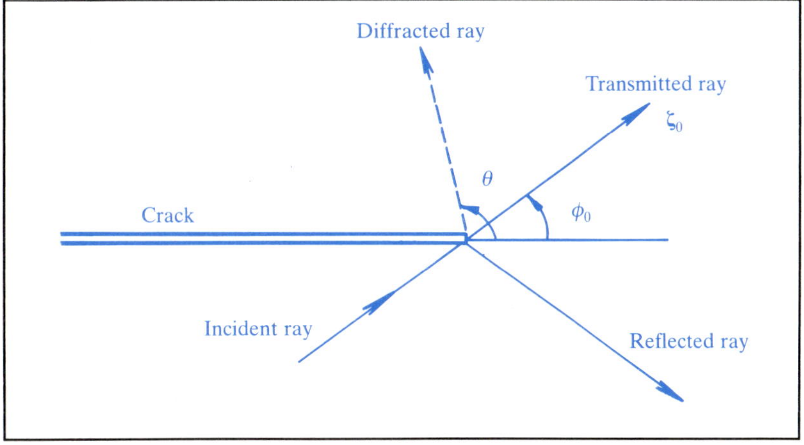

the Rayleigh wave speed c_R. In what follows, wave slownesses are used rather than wave speeds: $s_i = 1/c_i$.

There are two contributions to the far-field scattered signals, which comprise the geometrical result (high-frequency limit) and a first order correction in terms of an expansion in inverse powers of frequency. The results are most easily expressed in terms of components symmetric and antisymmetric with respect to the x axis. The elastic potentials may be written

$$\phi_\pm(x,y) = \int_{-\infty}^{\infty} \mathrm{P}_\pm(\zeta) e^{i\omega(\zeta x + \sqrt{s_p^2 - \zeta^2}\, y)}\, d\zeta \tag{3.225}$$

$$\psi_\pm(x,y) = \int_{-\infty}^{\infty} Q_\pm(\zeta) e^{i\omega(\zeta x + \sqrt{s_s^2 - \zeta^2}\, y)}\, d\zeta, \tag{3.226}$$

where

$$\mathrm{P}_+(\zeta) = (\tfrac{1}{2}s_s^2 - \zeta^2) R_+(\zeta) \tag{3.227}$$

$$\mathrm{P}_-(\zeta) = -\zeta (s_s^2 - \zeta^2)^{\frac{1}{2}} R_-(\zeta) \tag{3.228}$$

$$Q_+(\zeta) = \zeta (s_p^2 - \zeta^2)^{\frac{1}{2}} R_+(\zeta) \tag{3.229}$$

$$Q_-(\zeta) = (\tfrac{1}{2}s_s^2 - \zeta^2) R_-(\zeta). \tag{3.230}$$

The incident waves are also expressed in terms of their symmetric and antisymmetric components. One may write the incident amplitudes for compression waves as

$$\phi_0(x,y) = e^{i\omega(\zeta_0 x + \sqrt{s_p^2 - \zeta_0^2}\, y)} \tag{3.231}$$

$$\psi_0(x,y) = 0 \tag{3.232}$$

with Fourier transforms

$$A_+(\zeta_0) = -(\tfrac{1}{2}s_s^2 - \zeta_0^2) \tag{3.233}$$

$$A_-(\zeta_0) = \zeta_0(s_p^2 - \zeta_0^2)^{\frac{1}{2}}. \tag{3.234}$$

For an incident shear wave the incident amplitudes are

$$\phi_0(x,y) = 0 \tag{3.235}$$

$$\psi_0(x,y) = e^{i\omega(\zeta_0 x + \sqrt{s_s^2 - \zeta_0^2}\, y)} \tag{3.236}$$

with Fourier transforms

$$A_+(\zeta_0) = -\zeta_0(s_s^2 - \zeta_0^2)^{\frac{1}{2}} \tag{3.237}$$

$$A_-(\zeta_0) = -(\tfrac{1}{2}s_s^2 - \zeta_0^2). \tag{3.238}$$

The boundary conditions then give, beyond the crack tip ($x>0$) in the plane $y=0$,

$$\int_{-\infty}^{\infty} R_+(\zeta)e^{i\omega\zeta x}(s_p^2 - \zeta^2)^{\frac{1}{2}} d\zeta = 0 \tag{3.239}$$

and

$$\int_{-\infty}^{\infty} R_-(\zeta)e^{i\omega\zeta x}(s_s^2 - \zeta^2)^{\frac{1}{2}} d\zeta = 0. \tag{3.240}$$

The stress-free boundary conditions on the crack ($x \leqslant 0$) require

$$\int_{-\infty}^{\infty} R_\pm(\zeta)\{(\tfrac{1}{2}s_s^2 - \zeta^2) + \zeta^2(s_p^2 - \zeta^2)^{\frac{1}{2}}(s_s^2 - \zeta^2)^{\frac{1}{2}}\}e^{i\omega\zeta x} d\zeta = A_\pm(\zeta_0)e^{i\omega\zeta_0 x}. \tag{3.241}$$

The mathematical method is based on the Wiener–Hopf technique. A very clear exposition of this analytical scheme is given by Achenbach (1973, pages 365–372). The point is to notice that equations (3.239) to (3.241) imply that the integrands in equations (3.239) and (3.240) have no singularities in the upper half of the ζ plane, whereas the integrand in equation (3.241) has no singularities in the lower half plane (as these are the half planes in which the contours would be closed for the integrals to converge for $x<0$ and $x>0$ respectively). The method hinges on factorization into two functions, each of which is regular in one half plane and has singularities in the other half plane. This factorization is the key step, and the most difficult part, of the Wiener–Hopf procedure.

The reflection factors $R_{\pm}$ turn out to be

$$R_+(\zeta) = -\frac{1}{\pi i}\frac{1}{\zeta-\zeta_0}\left(\frac{s_p-\zeta_0}{s_p-\zeta}\right)^{\frac{1}{2}} \times \frac{1}{(s_R+\zeta)(s_R-\zeta_0)(s_s^2-s_p^2)}\frac{K_-(\zeta)}{K_-(\zeta_0)}\frac{A_+(\zeta_0)}{K(\zeta)} \tag{3.242}$$

and

$$R_-(\zeta) = -\frac{1}{\pi i}\frac{1}{\zeta-\zeta_0}\left(\frac{s_s-\zeta_0}{s_s-\zeta}\right)^{\frac{1}{2}} \times \frac{1}{(s_R+\zeta)(s_R-\zeta_0)(s_s^2-s_p^2)}\frac{K_-(\zeta)}{K_-(\zeta_0)}\frac{A_-(\zeta_0)}{K(\zeta)}. \tag{3.243}$$

Here the functions $A_{\pm}$ define the incident wave, with a projected wave slowness along the x axis of ζ_0. That is, the incident wavevector in the (x,y) plane has x–component $\omega\zeta_0$. Then, if the incident wave is a shear wave, we have

$$A_+(\zeta_0) = -\zeta_0(s_s^2-\zeta_0^2)^{\frac{1}{2}} \tag{3.244}$$

$$A_-(\zeta_0) = -(\tfrac{1}{2}s_s^2-\zeta_0^2). \tag{3.245}$$

The functions K, K_+ and K_- arise from the integral in the Wiener–Hopf method.

$$K(\zeta) = \frac{(s_s^2-2\zeta^2)^2+4\zeta^2(s_s^2-\zeta^2)^{\frac{1}{2}}(s_p^2-\zeta^2)^{\frac{1}{2}}}{2(s_s^2-s_p^2)(s_R^2-\zeta^2)} \tag{3.246}$$

$$\ln(K_{\pm}(\zeta)) = -\frac{1}{\pi}\int_{s_p}^{s_s}\tan^{-1}\left\{\frac{4z^2(z^2-s_p^2)^{\frac{1}{2}}(s_s^2-z^2)^{\frac{1}{2}}}{(s_s^2-2z^2)^2}\right\}\frac{dz}{z\pm\zeta}. \tag{3.247}$$

The details of the calculation of these functions are given in Achenbach and Gautesen (1977).

The diffracted fields are obtained from the asymptotic evaluation of the integrals in equations (3.225) and (3.226), giving cylindrical waves from the crack tip which are most readily written in cylindrical coordinates with

$$\zeta \equiv s\cos\phi \tag{3.248}$$

$$x \equiv r\cos\theta \tag{3.249}$$

as

$$\phi_+(r,\theta) = \frac{e^{i\omega r s_p}}{\sqrt{\omega r s_p}} s_p \sin\theta \, P_+(s_p \cos\theta)\sqrt{2\pi} e^{-\frac{1}{4}i\pi} \tag{3.250}$$

$$\phi_-(r,\theta) = \frac{e^{i\omega r s_p}}{\sqrt{\omega r s_p}} s_p \sin\theta \, P_-(s_p \cos\theta)\sqrt{2\pi} e^{-\frac{1}{4}i\pi} \tag{3.251}$$

$$\psi_+(r,\theta) = \frac{e^{i\omega r s_s}}{\sqrt{\omega r s_s}} s_s \sin\theta \, Q_+(s_s \cos\theta)\sqrt{2\pi} e^{-\frac{1}{4}i\pi} \tag{3.252}$$

$$\psi_-(r,\theta) = \frac{e^{i\omega r s_s}}{\sqrt{\omega r s_s}} s_s \sin\theta \, Q_-(s_s \cos\theta)\sqrt{2\pi} e^{-\frac{1}{4}i\pi}. \tag{3.253}$$

The effect of poles in the integrands is to give the geometrical part of the wavefield. This follows from the imposition of straight-line boundaries to the shadow zones and the reflection coefficient calculated as in Chapter 2 above.

Figure 35

Amplitudes of diffracted compression (P) and shear (SV) waves, for shear waves incident at 130° on a semi-infinite crack.

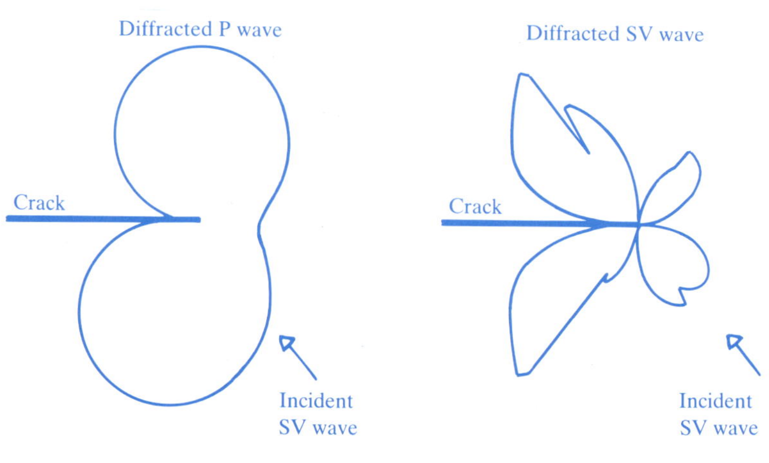

Figure 35 shows the angular distribution of waves scattered from a crack tip when a shear wave is obliquely incident. The singularity in the scattering at $\zeta = \zeta_0$ has been removed by multiplying the amplitudes by $(\zeta - \zeta_0)$. In applying the theory, this factor should be removed, but the results should not be used for diffractions close to a shadow boundary.

3.3.2 Practical aspects of geometric ray theory

The application of geometric ray theory to practical problems is a fairly straightforward extension of simple reflection theory. One requires a method of solving the geometrical problem, that is, of tracing rays from a transducer to a scatterer and from the scatterer to a detector. One also needs values for the diffraction coefficients, which relate the outgoing amplitudes to the incident ones. As each incident ray on an edge gives rise to many diffracted rays, it is necessary to impose some limit on the number of rays followed.

Figure 36

The cones of diffracted shear and compression waves produced when a shear wave impinges obliquely on a crack edge.

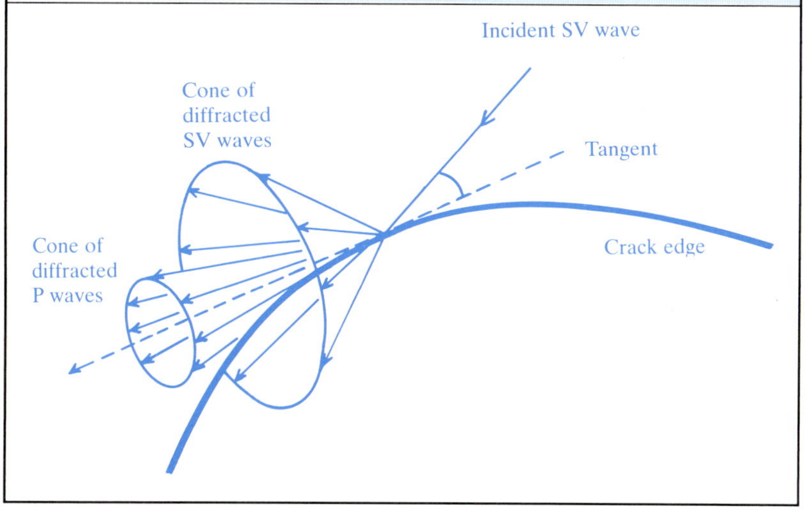

Achenbach and Gautesen (1977) have given expressions similar to those of the previous section for the diffraction of longitudinal waves from a crack edge. They consider arbitrary angles of incidence, whereas the results just presented apply only to cases where the incident wavevector lies in a plane normal to the crack. In the general case, as shown in Figure 36, an incident wave will produce two cones of diffracted bulk rays. The semi-angle of one cone is equal to the angle between the incident ray and the tangent to the crack edge, and the semi-angle of the mode converted cone is given by Snell's law. The theory has been extended (Gautesen, Achenbach and McMaken 1978) to include the surface waves which will be generated on the crack faces. These are important because strong resonance effects can be seen on finite cracks as a result of surface waves which are multiply reflected from the crack edges, shedding some energy into bulk waves at each reflection. The bulk waves which arise in this way are known as secondary diffractions, as opposed to the primary diffractions of bulk waves meeting an edge and producing further bulk waves.

In general we should include the uniform asymptotic correction to the field at the shadow boundaries. The problem is that the first order diffracted term has a singularity at that point, and the geometric solution has a step discontinuity. This situation has been considered by Lewis and Boersma (1969) in the context of electromagnetic scattering. They find that the shadow boundary is smoothed by replacing the step function multiplying the incident ray by two terms: a function $F(\eta)$ multiplying the incident or reflected ray, and a function $G(\eta)$ multiplying a cylindrical wave of the same type and amplitude. The functions F and G are defined in terms of the phase difference parameter η

$$\eta \equiv \sqrt{2\omega s_i r}\ \sin\tfrac{1}{2}(\theta - \phi_0) \tag{3.254}$$

where ϕ_0 is the incident angle of the wave of type i:

$$F(\eta) = \frac{e^{-\frac{1}{4}i\pi}}{\sqrt{\pi}} \int_{-\infty}^{\eta} e^{is^2}\, ds \tag{3.255}$$

$$G(\eta) = \frac{e^{\frac{1}{4}i\pi}}{2\eta\sqrt{\pi}}. \tag{3.256}$$

This is an example of a uniform asymptotic correction. Mathematically, the problem arises because, in the stationary phase evaluation of the contour integrals in equations (3.225) and (3.226), the stationary point lies at or near a singularity of the pre-exponential factor. Integrals of this type have been studied for general singularities and zeros by Bleistein (1966), who showed that they could be expanded in terms of a Weber function and its derivatives, each multiplied by a separate asymptotic series. It is interesting to note that the natural expansion functions for this half-plane diffraction problem would be Weber functions, arising from the use of parabolic coordinates. In the present case, the singularity is a simple pole, and the Weber function of order −1 which Bleistein's treatment gives is related to the Fresnel integral. The results then reduce to those of Lewis and Boersma (1969) used above. Gautesen (1980) and Achenbach, Gautesen and McMaken (1982) consider the uniform asymptotic theory for elastodynamics in greater detail.

From the special values of F

$$F(-\infty) = 0 \qquad F(0) = 0.5 \qquad F(\infty) = 1 \tag{3.257}$$

it is clear that F represents the smoothed step function in the geometric field at the shadow boundary. G removes the singularity in the diffracted part. Figure 37 shows the effect of the correction at a distance R from the crack tip such that $kR = 30$ and 300. At the larger range, the correction is negligible except very close to the shadow boundary.

As well as at shadow boundaries, geometric ray theory breaks down

Figure 37

Amplitudes of diffracted shear waves from a shear wave normally incident on a semi-infinite crack, at ranges (a) $kR = 30$, (b) $kR = 300$, showing the effect of the uniform asymptotic correction. Here k is the shear wavenumber, R is the distance from the crack tip.

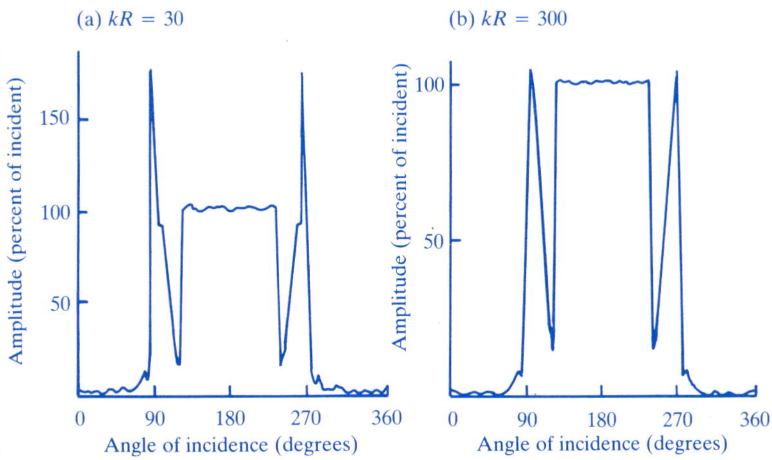

at caustics. These are points where, because of the curvature of the scatterer, rays from neighbouring points on the scatterer appear to intersect. At such points the cross-sectional area of the tube of rays shrinks to zero, so the usual argument about energy conservation in a ray bundle breaks down. This manifests itself mathematically as a singularity in the geometrical term in an integral to be asymptotically evaluated, in contrast to the breakdown at the shadow boundary which gives a singularity in the diffraction term. In fact, a diffracting edge is itself a caustic of the cone of rays diffracted from it, which again shows why geometric ray theory is unable to handle diffraction effects unless it is modified in the manner suggested by Keller (1953). Caustics may be classified by applying catastrophe theory (Thom 1975; Poston and Stewart 1978), and this has been done by Berry (1976).

The applications of caustics in nondestructive testing have been described by Doyle (1980) and Doyle, Latimer and Adler (1982), who suggest that they may provide useful information in addition to that obtained from ultrasonic imaging. In particular, they should be free of surface roughness and crack closure effects. Solutions near caustics again require uniform asymptotic treatments: the methods introduced by Bleistein (1966) are again appropriate. Keller (1957) has described the treatment of caustics in the geometrical theory of scalar waves, and Achenbach, Gautesen and McMaken (1978) have suggested that their hybrid method (described above in connection with the Kirchhoff approximation) may be used to avoid the difficulties which arise at shadow boundaries and caustics. Caustics are handled in a fairly straightforward way in Floyd's (1984, 1986) rigorous ray tracing scheme, which uses a Hamilton–Jacobi equation derived from the Helmholtz equation in a manner that includes terms dropped from the classical ray theory. McMaken's (1984) uniform theory, which reduces a surface integral over a crack face to a line integral along the crack edge, embraces caustics automatically.

Figure 38

The exact (triangles) and geometric ray theory (solid line) calculations of the scattering of an normally incident compression wave by a penny-shaped crack, as a function of size *ka*. The scattering angle is 30°, and the components of compression and shear displacement are shown, parallel and perpendicular to the incident wavevector and displacement (after Gubernatis, Domany and Krumhansl 1979).

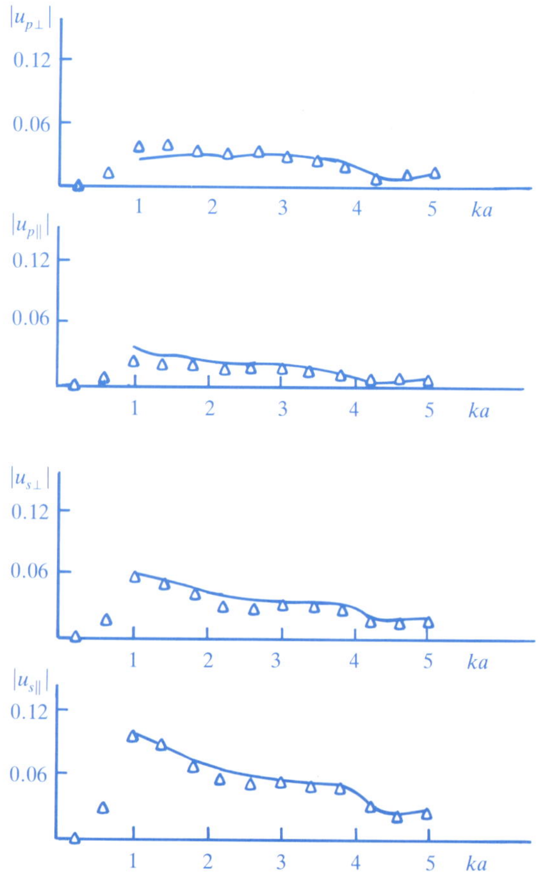

Geometric ray theory is an asymptotic theory, and is therefore expected to hold only for scattering from objects which are large compared with the wavelength. The range of validity may be estimated from Gubernatis, Domany and Krumhansl's (1979) study of the scattering from a penny-shaped crack. They compared the geometric ray theory result for scattering from a circular crack of radius a embedded in a solid with the exact result calculated by Mal (1968, 1970) by the numerical solution of a singular integral equation: see also reviews by Kraut(1976) and Martin and Wickham (1983). Figure 38 shows the scattering of a normally incident compression wave by such a crack. The scattering angle is 30°, and the components of compression and shear displacement are shown, parallel and perpendicular to the incident wavevector (which is also the direction of the incident displacement). The agreement is quite good for all $ka<1$, which suggests that the geometric theory and the quasistatic approximation together should cover the whole range of frequencies.

Figure 39

The amplitude of the back-scattered shear wave from a smooth strip-like crack of width $2a$, for $k_s a = 10$, as a function of angle of incidence. The exact solution is shown as a solid line, the geometric ray result as a chained line. After Chapman (1984).

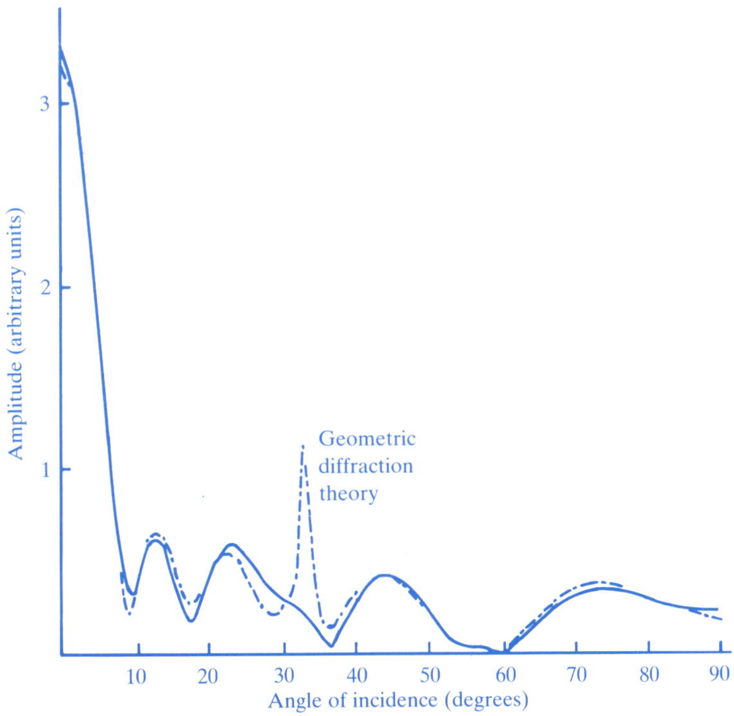

A similar conclusion regarding the range of validity of ray theory is reached by Chapman (1984), based on calculations of the back-scattered shear wave from a smooth strip-like crack of width $2a$, for $k_s a = 10$, at several angles of incidence. The exact solution calculated by Keogh (1983, 1986) is shown in Figure 39 as a solid line, the geometric ray result as a chained line. Comparing Figure 39 with Figure 27, one can see that the Kirchhoff and ray theory results can also be used to span the whole range of frequencies.

Langenberg and Schmitz (1986) point out that in the time domain the Kirchhoff approximation predicts two scattered pulses in the far field, one from each edge of the strip, with equal and opposite amplitudes. These are replicas of the incident pulse: in the specular direction, they coincide and give a pulse which is the time derivative of the incident pulse. Geometric ray theory predicts pulses of different amplitudes from the two edges, which is in better agreement with the exact solution.

It should not be assumed that ray theory is restricted to the treatment of crack-like defects. Aleshin and Mogilner (1982, 1984) have analysed the scattering by a cylinder in terms of a ray model. Such a treatment requires a careful treatment of the asymptotic forms of the relevant Bessel functions which occur in the exact solution (Section 3.1.3). We have already noted that the Sommerfeld–Watson transform leads to results which can be expressed in terms of ray theory. Using such analyses it is possible to derive diffraction coefficients to describe the scattering into bulk waves of creeping waves on the surface of a cylindrical flaw. Brind, Achenbach and Gautesen (1984) have also found a ray model useful in the analysis of forward scattering by a cylinder. Vopilkin (1985a, b) gives a number of applications of ray theory to both volumetric defects and cracks.

So far, we have only considered geometric ray theory as applied to isolated scatterers. Often the most important defects to be detected by nondestructive testing are surface-breaking. The application of ray theory to such cracks has been undertaken by Achenbach, Gautesen and Mendelson (1980). They based their analysis on the work of Mendelson, Achenbach and Keer (1980), which involved the numerical solution of two singular integral equations. Achenbach, Gautesen and Mendelson (1980) considered the scattering of surface waves. Their difficulties centred on computing the reflection and transmission coefficients of surface waves from the 90° corner at the root of the crack, where it breaks the open surface. No simple solution exists for this problem, so they relied on numerical results from Mendelson, Achenbach and Keer (1980), supplemented by finite difference and experimental results from Bond (1979). Similar problems will arise in the scattering of bulk waves from surface-breaking defects, suggesting that it might be advantageous to adopt a purely numerical method from the start of the problem for such cases.

3.4 Variational techniques

The calculus of variations is a familiar technique in mathematical physics. A particularly clear discussion is given by Courant and Hilbert (1961), and numerous applications are to be found in Forsyth's (1927) monograph. The use of the variational principle in scattering dates back to Hulthén (1944, 1947, 1948) and Schwinger (1947, 1950), who used slightly different formalisms. The methods were applied to electromagnetic problems by Levine and Schwinger (1948, 1949).

The variational technique is probably most familiar in the form of the Euler–Lagrange equations, which are the differential equations obtained from a minimum principle applied to a functional. In elastodynamics, for example, one can form a Lagrangian, apply Hamilton's variational principle, and obtain variational equations of motion. The equations are an expression of the principle of virtual work. These equations are entirely equivalent to the usual differential equations. The details are given by Achenbach (1973, page 49). One common application of the elastodynamic variational principle is in the calculation of the modes of vibration of rods and beams of complicated profiles. The variational principle is closely related to finite element methods of computation.

The variational method is a powerful way of obtaining approximate solutions to differential equations, including eigenvalue problems. For many problems, it is possible to find a functional which will, when the variational principle is applied, result in the original differential equation. In the context of eigenvalue problems of the form

$$\mathbf{L}\Lambda_n(x) = \lambda_n \mathbf{M}\Lambda_n(x) \tag{3.258}$$

with discrete eigenvalues λ_n and corresponding eigenfunctions Λ_n, the method centres on the functional ρ of a function ϕ:

$$\rho[\phi] = \frac{\int \phi\mathbf{L}\phi\, dx}{\int \phi\mathbf{M}\phi dx}. \tag{3.259}$$

For simplicity, we assume that the operators **L** and **M** are self-adjoint. If ϕ is an eigenfunction, it is clear that $\rho[\phi]$ is equal to the corresponding eigenvalue. If ϕ is an approximation to an eigenfunction, so that

$$\phi(x) = \Lambda_n(x)+\varepsilon\psi(x) \tag{3.260}$$

where ε is small, then

$$\rho[\phi] = \lambda_n+\varepsilon^2 \frac{\int \psi\mathbf{L}\psi\, dx-\lambda_n\int \psi\mathbf{M}\psi\, dx}{\int \Lambda_n\mathbf{M}\Lambda_n dx+2\varepsilon\int \Lambda_n\mathbf{M}\psi\, dx+\varepsilon^2\int \psi\mathbf{M}\psi\, dx}. \tag{3.261}$$

That is, if we take $\rho[\phi]$ as an estimate of λ_n then the error is of order ε^2 whereas the error in the estimate of the eigenfunction was of order ε.

To apply this result to scattering, one has to express the inhomogeneous scattering equation of the form

$$\mathbf{L}\phi(x) = I\phi_0(x) \tag{3.262}$$

as a homogeneous equation: here ϕ_0 represents the source term, with a strength I. This is done by introducing the scattering amplitude f, as in Section 3.2.1, and defining the scattering coefficient R by

$$R = \frac{f(\phi)}{I}. \tag{3.263}$$

Note that the functional dependence of f on ϕ has been made explicit here. Now the scattering equation (3.262) may be written as

$$\mathbf{L}\phi(x) = \frac{1}{R}\phi_0(x)f(\phi), \tag{3.264}$$

leading to the variational principle

$$\frac{1}{R} = \frac{\int \phi\phi_0 f(\phi)dx}{\int \phi\mathbf{L}\phi dx} \tag{3.265}$$

Weaver (1985) has derived a similar, but rather more general, form of this variational principle and applied it to wave propagation in random media.

To proceed further, we take the asymptotic forms of the scattering amplitudes. To simplify the expressions which follow, we shall treat scalar waves with wavevector $\mathbf{k}_0$ incident on a rigid body, for which ϕ is zero on the surface. Then the scattering amplitude is no longer a vector as in Section 3.2.1, and we may write

$$\phi(\mathbf{r}) \sim Ie^{i\mathbf{k}_0\cdot\mathbf{r}} + f(\phi)\frac{e^{ikr}}{kr} \tag{3.266}$$

where the scattering amplitude is

$$f(\phi) = -\frac{1}{4\pi}\int_{S'} \hat{\mathbf{n}}\cdot\nabla\phi(\mathbf{r}')e^{-ik\hat{\mathbf{r}}\cdot\mathbf{r}'}dS'. \tag{3.267}$$

As in Section 3.2.1, $\hat{\mathbf{r}}$ is a unit vector giving the direction from the scatterer to the point of observation, and $\hat{\mathbf{n}}$ is a unit outward normal to the scattering surface, S'. Now we use the asymptotic form to eliminate the source strength I from equation (3.262), to obtain

$$\mathbf{L}\phi(\mathbf{r}) = \frac{1}{R}f(\phi)e^{i\mathbf{k}_0\cdot\mathbf{r}}. \tag{3.268}$$

Now in this scalar wave case the operator **L** is defined by the exact solution of the Helmholtz scattering equation, that is, by the integral over the scattering surface of the Green function multiplied by the normal derivative of the wavefunction:

$$\begin{aligned}\phi(\mathbf{r}) &= I\phi_0(\mathbf{r}) - \int_{S'} \hat{\mathbf{n}}\nabla\phi(\mathbf{r}')\left[\frac{e^{ik|\mathbf{r}-\mathbf{r}'|}}{4\pi|\mathbf{r}-\mathbf{r}'|}\right]dS' \\ &= I\phi_0(\mathbf{r}) - \int_{S'} \hat{\mathbf{n}}\nabla\phi(\mathbf{r}')G(\mathbf{r}|\mathbf{r}')dS'\cdot\end{aligned} \qquad (3.269)$$

If we use this result in equation (3.268) and evaluate the field on the boundary surface of the scatterer, remembering the boundary condition on that surface, we find

$$\frac{1}{R}e^{i\mathbf{k}_0\cdot\mathbf{r}}\int_{S'}\hat{\mathbf{n}}\cdot\nabla\phi(\mathbf{r}')e^{-ik\hat{\mathbf{r}}\cdot\mathbf{r}'}dS' = -\int_{S'}\hat{\mathbf{n}}\nabla\phi(\mathbf{r}')\left[\frac{e^{ik|\mathbf{r}-\mathbf{r}'|}}{4\pi|\mathbf{r}-\mathbf{r}'|}\right]dS'\cdot \qquad (3.270)$$

This expression may be multiplied by the normal derivative of the adjoint solution $\hat{\mathbf{n}}\cdot\nabla\bar{\phi}$ and integrated over the scattering surface to give an expression for the scattering coefficient R:

$$R = -\frac{N_1 N_2}{D} \qquad (3.271)$$

where

$$N_1 = \int_{S'}\hat{\mathbf{n}}\cdot\nabla\bar{\phi}(\mathbf{r}')e^{-ik\hat{\mathbf{r}}\cdot\mathbf{r}'}dS', \qquad (3.272)$$

$$N_2 = \int_{S'}\hat{\mathbf{n}}\cdot\nabla\phi(\mathbf{r}')e^{i\mathbf{k}_0\cdot\mathbf{r}'}dS', \qquad (3.273)$$

$$D = \int_S\int_{S'}\hat{\mathbf{n}}\cdot\nabla\bar{\phi}(\mathbf{r})\left[\frac{e^{ik|\mathbf{r}-\mathbf{r}'|}}{4\pi|\mathbf{r}-\mathbf{r}'|}\right]\hat{\mathbf{n}}'\cdot\nabla\phi(\mathbf{r}')dSdS'. \qquad (3.274)$$

This is the required variational principle. With a specified approximate field ϕ the adjoint field $\bar{\phi}$ is known: it is the solution of the reciprocal scattering problem, with $\mathbf{k}_0$ replaced by $-k\hat{\mathbf{r}}$ and $k\hat{\mathbf{r}}$ replaced by $-\mathbf{k}_0$ (Morse and Feshbach 1953, page 1135). Thus equation (3.274) can be used to estimate the far field scattering coefficient R. Ogilvy (1986) has used this form of the variational principle to assess the accuracy of the Kirchhoff approximation in scattering from rough surfaces.

Many other formulations of the variational principle are possible. Some are based on minimum energy principles, and are analogous to the treatment of the elastostatic case by Hashin and Shtrikman (1962a,b, 1963) which resulted in the Hashin–Shtrikman bounds on the elastic constants of composite materials. Generally, variational methods have been of most use in elastodynamics in studies of wave propagation in granular media and in scattering from rough surfaces: these topics are discussed further in Sections 4.6.1 and 4.6.2.

3.5 Numerical modelling methods

In the study of the scattering of elastic waves, in practical applications such as nondestructive testing, one frequently wants to treat complicated geometrical situations, which are not amenable to analytic treatment. A further difficulty is that one often works in the region where the size of the scatterer is comparable with the sound wavelength. At worst this makes the problem intractable by analytic means, and at best it necessitates the use of long and involved expansions in terms of standard functions which may or may not converge rapidly. We have just seen that even relatively simple scatterers can lead to difficult analysis and solutions which require numerical evaluation when they are associated with a surface. When one adds to this the complications of handling a range of frequencies making up a pulse, and the need to handle incident wavefronts of finite width rather than infinite plane waves, the amount of computation involved is formidable. Even expert practitioners of analytic methods recognize the value of this (Scandrett, Kriegsmann and Achenbach 1986a,b; Scandrett and Achenbach 1987).

Rather than face both tedious analysis and lengthy numerical work, one can opt for a completely numerical approach. The commonest methods are based on finite differences and finite elements. There are distinctions between the approaches and the solution techniques used in these two methods, but they are quite closely related and equivalences between the two schemes can be drawn up (Morton 1976). Historically, finite difference methods have precedence. They were introduced by Euler (1768) for one-dimensional problems, and the extension to two dimensions was made by Runge (1908). Finite element methods, on the other hand, were not introduced until electronic digital computers became available. The term *finite element method* was first used by Clough (1960), although the foundations of the method were laid by Turner *et al.* (1956) and Argyris (1960).

Although finite difference methods for elastic wave problems have been in use for some time (Alterman 1968; Bertholf 1967), their application to nondestructive testing problems did not begin until about 1973. At about that time Harumi, Suzuki and Saito (1973) modelled the near field of a transducer, and Rose and Meyer (1975) used a numerical model to confirm the results of their model calculation of transducer fields. Calculations of scattering have been undertaken since about 1976, and the first published work was that of Harumi, Saito and Fujimori (1979), Saito *et al.* (1982), Blake, Bond and Downie (1982) and Bond and Blake (1983). A review is given by Bond (1982), and an impressive compilation of results may be found in Harumi (1986). The finite element method has not been as widely applied to nondestructive testing wave propagation problems as the finite difference methods, although for static stress problems it is generally the preferred technique. In seismology the method is well established (Lysmer and Drake 1972), although there a single frequency rather than a general time dependence is often assumed. Smith (1975) described finite element analysis for body wave propagation. A detailed analysis of the application of time-dependent finite elements to the diffraction by a crack tip was given by Bažant, Glazik and Achenbach (1976). Dewey *et al.* (1983) have used the same approach for modelling ultrasonic inspection of welds. Ootsuki and Harumi (1983) have combined a lumped-mass finite element model with their particle model, using the finite element model near defects. Tsao, Varadan and Varadan (1983) have developed a hybrid finite element and wave expansion technique for subsurface flaws, and the method has also recently been used by Datta and Shah (1982) in conjunction with an analytical technique for the far field. Finite element analyses of the scattering of Lamb waves have been made by Koshiba *et al.* (1983) and Koshiba, Karakida and Suzuki (1984). To date, there seem to have been no finite element calculations on elastic wave scattering problems which exploit what is usually claimed to be the major advantage of finite elements over finite differences, namely the ease with which complicated geometries can be treated. In what follows we shall concern ourselves only with finite difference methods.

3.5.1 Finite difference equations for homogeneous media

The basic idea of the finite difference method is very simple. The differential form of the wave equation for a continuum is replaced by a discrete set of equations in which the values of the field quantities (in the present case, the displacements) are defined only on the points of intersection of a grid. The partial derivatives in the wave equation are approximated, to a given accuracy, by combining Taylor expansions of functions defined over the grid. The method we describe here works with the wave equation in its second order form. It is also possible (Clifton 1967) to express the problem in terms of coupled first order equations, using velocity and stress as the field quantities. Duncan (1985) has recently used such a scheme for elastic wave scattering calculations. Bayliss (1966) also used used velocity and stress fields, to produce a scheme which is accurate to second order in time and fourth order in space.

The wave equation for homogeneous, isotropic media is, in terms of the displacements $\mathbf{u}(\mathbf{x},t)$,

$$(\lambda+\mu)\nabla\nabla\cdot\mathbf{u}+\mu\nabla^2\mathbf{u} = \rho\frac{\partial^2\mathbf{u}}{\partial t^2} \tag{2.40}$$

where we have assumed that there are no body forces. If we consider a plane strain situation, in which the displacements are zero in the y

direction, so that $\mathbf{u} = (u,0,z)$, and there is no dependence on z, we may write

$$u_{,tt} = c_p^2 u_{,xx} + (c_p^2 - c_s^2) w_{,xz} + c_s^2 u_{,zz} \tag{3.275}$$

$$w_{,tt} = c_p^2 w_{,zz} + (c_p^2 - c_s^2) u_{,xz} + c_s^2 w_{,xx} . \tag{3.276}$$

If we now define the variables u and w on a uniform Cartesian mesh, of spacing h in both x and z directions, and at time intervals Δt, we may write

$$\mathbf{u}(m,n,l) \equiv \mathbf{u}((mh,nh),l\Delta t). \tag{3.277}$$

Then, for the second derivatives we use the second order finite difference representations given by Abramowitz and Stegun (1965, page 884):

$$u_{,tt}(m,n,l) = \frac{1}{(\Delta t)^2}(u(m,n,l+1) - 2u(m,n,l) + u(m,n,l-1)) \tag{3.278}$$

$$u_{,xx}(m,n,l) = \frac{1}{h^2}(u(m+1,n,l) - 2u(m,n,l) + u(m-1,n,l)) \tag{3.279}$$

$$\begin{aligned} u_{,xz}(m,n,l) = \frac{1}{4h^2}(&u(m+1,n+1,l) - u(m-1,n+1,l) \\ &- u(m+1,n-1,l) + u(m-1,n-1,l)) \cdot \end{aligned} \tag{3.280}$$

Substituting back, and defining the time−to−space step ratio $\sigma = \dfrac{\Delta t}{h}$, the finite difference equations become

$$\begin{aligned} u(m,n,l+1) = {} & 2(1 - c_p^2\sigma^2 - c_s^2\sigma^2)u(m,n,l) - u(m,n,l-1) \\ & + c_p^2\sigma^2(u(m+1,n,l) + u(m-1,n,l)) \\ & + c_s^2\sigma^2(u(m,n+1,l) + u(m,n-1,l)) \\ & + \tfrac{1}{4}\sigma^2(c_p^2 - c_s^2)(w(m+1,n+1,l) - w(m-1,n+1,l) \\ & \qquad - w(m+1,n-1,l) + w(m-1,n-1,l)) \end{aligned} \tag{3.281}$$

with a similar equation for w.

Harumi, Igarashi and Saito (1978) use a finite difference scheme based on a representation of the medium in terms of a square mesh of masses linked by vertical and horizontal springs and diagonal rods. This results in equations which differ slightly from (3.281):

$$\begin{aligned} u(m,n,l+1)^{\text{particle}} &= u(m,n,l+1)^{\text{finite diff}} \\ &\quad +\tfrac{1}{2}c_s^2\sigma^2(u(m+1,n+1,l)+u(m-1,n+1,l) \\ &\qquad +u(m+1,n-1,l)+u(m-1,n-1,l) \\ &\qquad -4u(m,n,l)) \\ &\quad -c_s^2\sigma^2(u(m+1,n,l)+u(m-1,n,l) \\ &\qquad +u(m,n+1,l)+u(m,n-1,l) \\ &\qquad -4u(m,n,l)). \end{aligned} \tag{3.282}$$

This extra term is simply $c_s^2\sigma^2$ times the difference between two equally plausible approximations to $\nabla^2 u$, and the schemes should therefore give similar results.

Before using a finite difference scheme, we should satisfy ourselves that it is stable, and we should consider the accuracy of the method, and we shall consider these aspects in the following sections.

3.5.2 Stability of finite difference method

The stability of a finite difference method is a property of the entire problem: that is, it depends on the finite difference representations both of the bulk medium *and* the boundary conditions. Commonly, however, a method of stability analysis is used which ignores the boundary conditions, and we shall use that method here: it was applied to the elastic wave system by Alterman and Loewenthal (1970). The method involves studying the propagation of a plane wave through an infinite grid: as it is a single-frequency wave the method is known as the Fourier method (Ames 1977, page 47), although it appears to have been first suggested by von Neumann during the Second World War (O'Brien, Hyman and Kaplan 1951).

Consider the plane wave

$$\mathbf{u}(m,n,l) = \mathbf{u}e^{i(k_x m+k_z n)h}\gamma^l \tag{3.283}$$

where the time dependence is contained within the factor γ, and the condition for stability is that $|\gamma|\leq 1$. To simplifiy the equations which follow, we define

$$\alpha \equiv c_p\Delta t/h \tag{3.284}$$

$$\beta \equiv c_s\Delta t/h \tag{3.285}$$

and substitute equation (3.283) into (3.281). Removing the exponential factor gives

$$\begin{aligned} u\gamma &= 2(1-\alpha^2-\beta^2)u\gamma^{-1} \\ &\quad +2\alpha^2 u\cos(kh\cos\theta)+2\beta^2 u\cos(kh\sin\theta) \\ &\quad +\tfrac{1}{4}(\alpha^2-\beta^2)w(\cos(kh(\cos\theta+\sin\theta))-\cos(kh(\cos\theta-\sin\theta))) \end{aligned} \tag{3.286}$$

and similarly for w. By evaluating the determinant of the coefficients of u and w for the worst case ($kh\ \sin\theta = \frac{1}{4}\pi$ and $kh\ \cos\theta = \frac{1}{4}\pi$) Alterman and Loewenthal (1970) were able to show that, for stability, one must have

$$\frac{\Delta t}{h} \leqslant (c_s^2 + c_p^2)^{-\frac{1}{2}}. \tag{3.287}$$

This is, roughly speaking, a requirement that a signal shall not be able to propagate across one mesh spacing in less than one time step.

Ilan and Loewenthal (1976) studied the effects of the boundary conditions on the stability of finite difference schemes. Their approach was two-fold: empirical, using trial calculations with the chosen finite difference method, and theoretical, using a modification of the matrix method. The basic idea of this approach is quite simple. We can arrange all the displacement values at the mesh points at time $l\Delta t$ as a vector $\mathbf{U}^l$, which has dimension $2MN$ for the two displacement directions, M points in the x direction and N points in the z direction:

$$\mathbf{U}^l \equiv \begin{Bmatrix} u(1,1,l) \\ w(1,1,l) \\ u(2,1,l) \\ \cdot \\ \cdot \\ u(M,N,l) \\ w(M,N,l) \end{Bmatrix}. \tag{3.288}$$

Then the finite difference scheme may be written as

$$\mathbf{U}^{l+1} = \mathbf{O}\mathbf{U}^l - \mathbf{1}\mathbf{U}^{l-1} \tag{3.289}$$

where $\mathbf{1}$ is the unit matrix of dimension $2MN \times 2MN$ and $\mathbf{O}$ is a matrix with the same dimensions which represents the coefficients of $u(m,n,l)$ and $w(m,n,l)$ in equation (3.281). We can modify the equations to make them resemble two-level rather than three-level schemes (the level here refers to the number of time steps involved) by defining the vector $\mathbf{V}^l$:

$$\mathbf{V}^l \equiv \begin{Bmatrix} \mathbf{U}^l \\ \mathbf{U}^{l-1} \end{Bmatrix}. \tag{3.290}$$

when we may write

$$\mathbf{V}^{l+1} = \mathbf{P}\mathbf{V}^l \tag{3.291}$$

with

$$\mathrm{P} = \begin{Bmatrix} \mathbf{O} & \mathbf{1} \\ \mathbf{1} & 0 \end{Bmatrix}. \tag{3.292}$$

Clearly now the operation of the finite difference scheme is equivalent to raising the matrix P to a high power, and in the limit P^l will be dominated by the highest eigenvalue of P, p_{max} say. If the scheme is to be stable, p_{max} must not have an absolute value greater than unity. This condition is most easily derived by considering the eigenvalues and eigenvectors of P. If the eigenvectors are $\mathbf{X}^{(n)}$ with eigenvalues ε_n, we may expand $\mathbf{V}^1$ as

$$\mathbf{V}^1 = \sum_n \chi_n \mathbf{X}^{(n)} \tag{3.293}$$

and then, as

$$\mathbf{P}\mathbf{X}^{(n)} = \varepsilon_n X^{(n)}, \tag{3.294}$$

we have

$$\mathbf{V}^{l+1} = \sum_n \chi_n \varepsilon_n^l \mathbf{X}^{(n)}. \tag{3.295}$$

From this it is clear that if all the eigenvectors of P have absolute values less than or equal to unity, no initial vector $\mathbf{V}^1$, and in particular no error vector, will grow without limit.

If boundary conditions are neglected, the matrix P is easily analysed (see Boore 1970, 1972) by taking advantage of the diagonally banded structure of the matrix. The result is the same stability as was derived above. When the boundary conditions are included, Ilan and Loewenthal (1976) had to make use of Gerschgorin's theorems to analyse the stability. The first theorem states that the eigenvalue of largest absolute magnitude of a square matrix cannot exceed the sum of the moduli of the elements in any row or column. The second theorem states that if S_s is the sum of the moduli of the elements along row i of the matrix, excluding the diagonal element ii, then each eigenvalue of the matrix lies inside or on the boundary of at least one of the circles centred on the diagonal element and of radius S_s. Their analysis was particularly useful in identifying regions of instability when their original difference scheme was applied to materials of low shear constants. Such considerations might be important for finite difference models which treated the coupling of solids with polymeric materials or viscous fluids, but should not arise for most engineering materials.

In practice, there seems to be little advantage in applying the matrix method explicitly. It is just as effective to test the finite difference scheme on a small problem, where the effect of calculating for a large number of timesteps is to raise the propagation matrix P to a high power, when the vector $\mathbf{V}^l$ will be dominated by the largest eigenvalue. If the matrix method is to be applied, it is best to derive the matrix explicitly for the particular scheme under consideration, using the computer code which will be used for the finite difference calculations, inserting unit disturbances at points in the mesh, and finding their effects at neighbouring points. This approach not only reduces the coding effort, it also minimizes the risk of transcription errors. If formal analysis is to be undertaken, modern computer codes for algebraic manipulation can be of great assistance: Liska (1984) has begun to produce codes for general problems.

3.5.3 Accuracy of finite difference method

The stability criterion we gave above only sets a limit on the ratio of the grid spacing to the time interval used in the finite difference method. It offers no guidance as to the grid spacing that is required if the solution is to be accurate rather than simply stable. If we concentrate on the problem of a single frequency, giving a sinusoidal wave, we can study the accuracy. Intuitively, it is clear that a sine wave cannot be described with fewer than four points for each oscillation, to describe its peaks, zeros and troughs. It seems reasonable to assume that about twice as many would be required to represent it with reasonable accuracy.

Consider a plane compression wave travelling through a grid. From the difference equations, writing $\gamma = e^{i\omega\Delta t}$, where ω will be real if the stability criterion is satisfied, and taking $(k_x, 0, k_z) = k(\cos\theta, 0, \sin\theta)$, we find

$$\sin^2\frac{\omega\Delta t}{2} = \alpha^2\sin^2\left(\frac{kh\cos\theta}{2}\right)+\beta^2\sin^2\left(\frac{kh\sin\theta}{2}\right) + \tfrac{1}{4}(\alpha^2-\beta^2)\tan\theta\sin(kh\cos\theta)\sin(kh\sin\theta), \tag{3.296}$$

and from this we can extract the phase velocity, $v = \omega/k$. For simplicity, consider a wave propagating parallel to the x axis, so that $\theta=0$. Then

$$\frac{v}{c_p} = \frac{\omega}{kc_p} = \frac{2}{kc_p\Delta t}\sin^{-1}\left(\alpha\sin\frac{kh}{2}\right) \tag{3.297}$$

or, if we define N to be the number of points taken per wavelength,

$$\frac{v}{c_p} = \frac{N}{\pi}\,\frac{h}{c_p\Delta t}\,\sin^{-1}\left(\frac{\Delta t}{h}c_p\sin\frac{\pi}{N}\right). \tag{3.298}$$

From this expression it is immediately apparent that as $N\to\infty$ the phase velocity in the finite difference model approaches the exact value.

The results in Figure 40 show that the error in the velocity is about one per cent when ten points are taken to each wavelength. In practical calculations, it is advisable to take rather more than ten points to represent the wavelength of the dominant frequency, so that the higher frequencies which will be included in any pulse are adequately represented.

Figure 40

The ratio of the phase velocity in the finite difference calculation to the exact value v/c_p is plotted as a function of the number of points per wavelength for various values of the stability parameter $\frac{\Delta t}{h}c_p$.

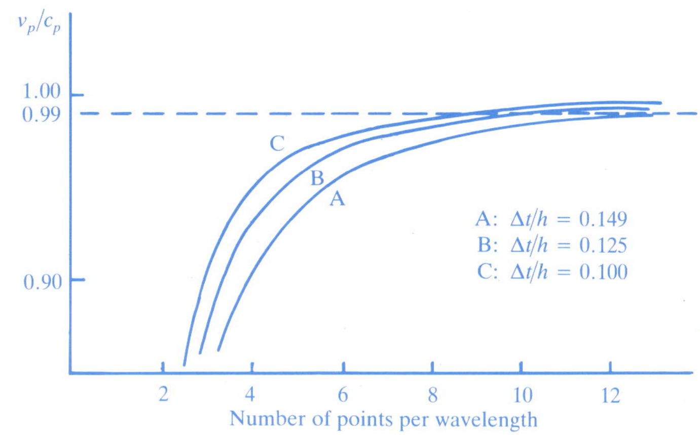

3.5.4 Finite difference scheme for inhomogeneous media

If we allow material constants to be functions of position, we can treat inclusions, complicated shapes, and transmission through boundaries. We have to return to the basic differential equation and write it in the appropriate form. In components, this gives:

$$u_{,tt} = \frac{\partial}{\partial x}(c_p^2 u_{,x} + (c_p^2 - 2c_s^2)w_{,z}) + \frac{\partial}{\partial z}c_s^2(w_{,x} + u_{,z}) \tag{3.299}$$

$$w_{,tt} = \frac{\partial}{\partial z}(c_p^2 w_{,z} + (c_p^2 - 2c_s^2)u_{,x}) + \frac{\partial}{\partial x}c_s^2(w_{,x} + u_{,z}). \tag{3.300}$$

If we retain the centred difference approach, we take

$$\begin{aligned}\frac{\partial}{\partial x}(c_p^2 u_{,x})(m,n,l) = \{&c_p^2(m+\tfrac{1}{2},n)(u(m+1,n,l)-u(m,n,l)) \\ &-c_p^2(m-\tfrac{1}{2},n)(u(m,n,l)-u(m-1,n,l))\}/h^2,\end{aligned} \tag{3.301}$$

where we have introduced

$$c_p^2(m+\tfrac{1}{2},n) = \tfrac{1}{2}(c_p^2(m+1,n)+c_p^2(m,n)). \tag{3.302}$$

The cross derivatives are

$$\frac{\partial}{\partial z}(c_p^2 u_{,x})(m,n,l) = \{(c_p^2 u_{,x})(m,n+1,l)-(c_p^2 u_{,x})(m,n-1,l)\}/(2h)$$
$$= \{c_p^2(m,n+1,l)(u(m+1,n+1,l)-u(m-1,n+1,l))$$
$$-c_p^2(m,n-1,l)(u(m+1,n-1,l)-u(m-1,n-1,l))\}/4h^2 \tag{3.303}$$

and

$$\frac{\partial}{\partial x}(f^2 w_{,z})(m,n,l) = \{(f^2 w_{,z})(m+1,n,l)-(f^2 w_{,z})(m-1,n,l)\}/2h$$
$$= \{f^2(m+1,n,l)(w(m+1,n+1,l)-w(m+1,n-1,l))$$
$$-f^2(m-1,n,l)(w(m-1,n+1,l)-w(m-1,n-1,l))\}/4h^2. \tag{3.304}$$

In the cross terms, $f^2 = c_p^2 - 2c_s^2$.

The final finite difference equation for u is then

$$\begin{aligned} u_{,tt}(m,n,l) = \{&2(c_p^2(m+1,n)+c_p^2(m,n))u(m+1,n,l) \\ &-2(c_p^2(m+1,n)+2c_p^2(m,n)+c_p^2(m-1,n))u(m,n,l) \\ &-2(c_s^2(m,n+1)+2c_s^2(m,n)+c_s^2(m,n-1))u(m,n,l) \\ &+2(c_p^2(m-1,n)+c_p^2(m,n))u(m-1,n,l) \\ &+2(c_s^2(m,n+1)+c_s^2(m,n))u(m,n+1,l) \\ &+2(c_s^2(m,n-1)+c_s^2(m,n))u(m,n-1,l) \\ &+f^2(m+1,n)(w(m+1,n+1,l)-w(m+1,n-1,l)) \\ &-f^2(m-1,n)(w(m-1,n+1,l)-w(m-1,n-1,l)) \\ &+c_s^2(m,n+1)(w(m+1,n+1)-w(m-1,n-1)) \\ &-c_s^2(m,n-1)(w(m+1,n-1)-w(m-1,n-1))\}/(4h^2). \end{aligned} \tag{3.305}$$

Again, the equation for w is similar. The use of a fully inhomogeneous method is very time-consuming, as the inhomogeneous scheme takes about three times as much computation as the homogeneous one. For simple geometries, it is possible to treat boundaries in a simpler manner.

3.5.5 Boundary conditions

Rigid boundaries are trivial, as they only require the setting of displacements to zero at each time step. For stress-free boundaries, the situation is rather more complicated. For example, if we take a plane perpendicular to the z axis we must have the stresses

$$(c_p^2 - 2c_s^2)u_{,x} + c_p^2 w_{,z} = \tau_{yy}/\rho = 0 \tag{3.306}$$

$$c_s^2(u_{,z} + w_{,x}) = \tau_{xy}/\rho = 0. \tag{3.307}$$

One way of achieving a stable scheme requires the introduction of a row of artificial points just outside the boundary. The displacements of these points are then determined at each time step so as to give zero stress on the boundary. Then one may use centred differences to form the derivatives parallel to the boundary, and one-sided differences for the derivatives perpendicular to the boundary. This was the method adopted by Alterman and Rotenberg (1969), when they found that the method of Alterman and Karal (1968) using centred differences in both directions was unstable for their case. Improved approximations were introduced by Ilan, Ungar and Alterman (1975), based on the method used by Lax and Wendroff (1960, 1964). In this 'composed' scheme the technique is to replace the derivatives perpendicular to the surface with centred differences parallel to the surface and differences in time, using the differential equation itself. Because the normal derivatives are substituted for in this way there is no need for the layer of artificial points outside the surface. The problems of stability and accuracy were investigated further by Ilan and Loewenthal (1976). They found that the mixed (centred and one-sided) scheme gave the best results for pulses which contained a wide range of frequencies, but that the centred and composed schemes were better when the input pulse was well smoothed. They went on to develop a modification of the composed scheme which was stable over the whole range of material properties (the centred scheme becomes unstable for Poisson's ratio greater than 0.45, the mixed scheme for Poisson's ratio less than 0.43, and the original composed scheme for Poisson's ratio greater than 0.26). For most structural materials, the mixed boundary scheme is adequate, and we develop the relevant equations below.

Using centred finite differences parallel to the surface and one-sided differences normal to the surface, the equations for zero stress become, on the plane $z=n_0h$,

$$\begin{aligned}(c_p^2-2c_s^2)(u(m+1,n_0,l)-u(m-1,n_0,l))\\ +2c_p^2(w(m,n_0+1,l)-w(m,n_0,l)) = 0\end{aligned} \tag{3.308}$$

and

$$2(u(m,n_0,l)-u(m,n_0-1,l))+w(m+1,n_0,l)-w(m-1,n_0,l) = 0. \tag{3.309}$$

For corners, the problem is a little more complicated. Consider a corner at $(x,z) = (0,0)$. The difference equations for the bulk fix the displacements at $(-1,0)$, $(0,0)$, $(0,-1)$ and $(-1,-1)$, and the plane boundary conditions given above fix the values at $(-1,1)$ and $(1,-1)$. To obtain the values at $(0,1)$ and $(1,0)$ the equations for the two intersecting boundary planes are solved simultaneously, so as to match the boundary conditions at the corner. The result is

$$w(1,0)=\tfrac{4}{3}(w(0,0)-\tfrac{1}{4}w(-1,0)-\tfrac{1}{2}(u(0,0)-u(0,-1))) \tag{3.310}$$

$$u(0,1)=\tfrac{4}{3}(u(0,0)-\tfrac{1}{4}u(0,-1)-\tfrac{1}{2}(w(0,0)-w(-1,0))) \tag{3.311}$$

and, if we define

$$q = (c_p^2-2c_s^2)/2c_p^2, \tag{3.312}$$

$$w(0,1)=(w(0,0)-q^2w(0,-1)-q(u(0,0)-u(-1,0)))/(1-q^2) \tag{3.313}$$

$$u(1,0)=(u(0,0)-q^2u(-1,0)-q(w(0,0)-w(0,-1)))/(1-q^2), \tag{3.314}$$

which leaves the point (1,1) undetermined. To fix the displacements there, we use a new finite difference scheme and impose the condition

$$(\tau_{xx}+\tau_{yy}) = \tau_{xy} = 0 \tag{3.315}$$

at the corner. This gives the two additional equations

$$u(1,1) = u(-1,-1)+w(-1,1)-w(1,-1) \tag{3.316}$$

$$w(1,1) = w(-1,-1)+u(1,-1)-u(-1,1). \tag{3.317}$$

The above scheme has been found empirically to be stable under the same conditions as the infinite domain equations are stable.

Figure 41

The finite difference formulation for a corner. Mesh points marked o are normal internal points, those marked × are ordinary boundary points, determined by equations (3.308), (3.309). The displacements of points marked A are found from equations (3.310) to (3.314), those of points marked B from equations (3.316) and (3.317).

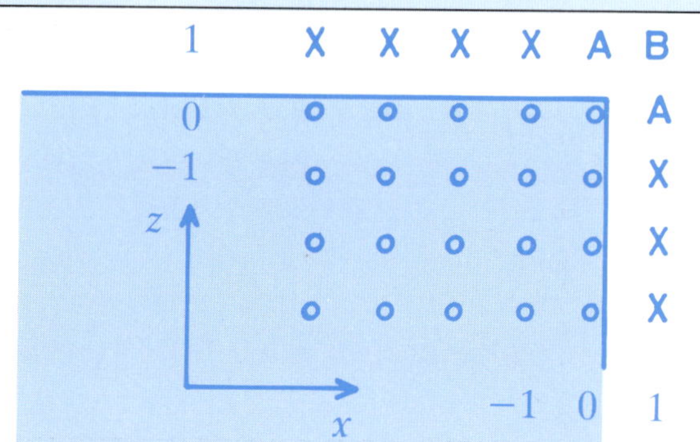

3.5.6 Cracks and laminations

Planar defects may be treated by a simple extension of the equations for planar surfaces given above. Again the rigid defects are trivial. A free-surfaced defect is treated as two coincident planes, each with zero stress. This gives a defect of negligible width, but assumes that the surfaces are not brought into contact by the displacements in the wave (see Section 4.6.3).

For the free-surfaced defect, then, we have to introduce two fictitious rows of points, one for each surface. Most of those planes are normal free surfaces, and the corners formed when a defect breaks through the surface may be treated in the same way as the corners described above. At the ends of defects in the bulk, the fictitious points introduced have to have their displacements set equal to those of points in the bulk adjacent to the defect, to establish the correct connectivity and end the defect smoothly. It should be noted that this scheme models a crack with zero width, whereas most formulations treat a crack as a slot a few mesh spacings wide.

Figure 42

The finite difference formulation for a surface breaking crack. Mesh points marked o are normal internal points, those marked × are ordinary boundary points, determined by equations (3.308), (3.309). The displacements of points marked A are found from equations (3.310) to (3.314), those of points marked B from equations (3.316) and (3.317). Points which represent the same physical point are joined by arrows.

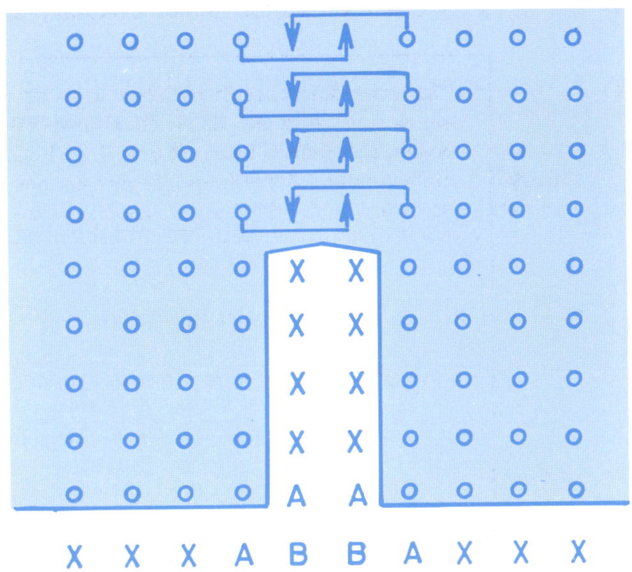

Similar simplifications are possible in layered systems, where the most efficient method is to apply the homogeneous scheme throughout the layers, and then to run along the boundaries between layers adjusting the displacements of the points on each side of the joining line only.

3.5.7 Higher order finite difference schemes

It is natural to ask whether computational economies are possible by adopting a finite difference scheme which is accurate to higher order. The potential advantage is that fewer mesh points would be needed for accurate representation of the waveform. If the number of mesh points in each direction is reduced by a factor f, in two dimensions the number of points is reduced by f^2. The time step, one may expect, may be increased whilst still maintaining the stability requirement that no disturbance shall propagate over a mesh spacing in a time step. The potential reduction in the total number of time steps multiplied by the number of points is then about f^3. It is likely, however, that this is offset by the increased complexity of the finite difference equations at each point.

The derivation of the fourth order equations follows the same pattern as for the second order scheme, as does the analysis of stability. The basic difference rules used were

$$\begin{aligned} f_{,xx}(m,n,l) = \frac{1}{12h^2}\{&-f(m+2,n,l)+16f(m+1,n,l) \\ &-30f(m,n,l)+16f(m-1,n,l)-f(m-2,n,l)\} \\ &+O(h^4) \end{aligned} \tag{3.318}$$

$$\begin{aligned} f_{,xy}(m,n,l) = {} & \{f(m+1,n+1,l)+f(m-1,n-1,l) \\ & -f(m+1,n-1,l)-f(m-1,n+1,l)\}/3h^2 \\ & -\{f(m+2,n+2,l)+f(m-2,n-2,l) \\ & -f(m+2,n-2,l)-f(m-2,n+2,l)\}/48h^2 \\ & +O(h^4). \end{aligned} \tag{3.319}$$

When these are substituted into the elastic wave equations, we obtain

$$\begin{aligned} u(m,n,l+1) = {} & 2u(m,n,l)-u(m,n,l-1)-\frac{5}{2}(c_s^2+c_p^2)\sigma^2 u(m,n,l) \\ & +\frac{1}{2}c_p^2\sigma^2(16u(m+1,n,l)+16u(m-1,n,l)-u(m+2,n,l)-u(m-2,n,l)) \\ & +c_s^2\sigma^2(16u(m,n+1,l)+16u(m,n-1,l)-u(m,n+2,l)-u(m,n-2,l)) \\ & +\frac{1}{3}\sigma^2(c_p^2-c_s^2)(w(m+1,n+1,l)+w(m-1,n-1,l) \\ & \quad -w(m-1,n+1,l)-w(m+1,n-1,l)) \\ & -\frac{1}{48}\sigma^2(c_p^2-c_s^2)(w(m+2,n+2,l)+w(m-2,n-2,l) \\ & \quad -w(m+2,n-2,l)-w(m-2,n+2,l)). \end{aligned} \tag{3.320}$$

The expression for w is similar.

The stability condition for the infinite medium using the fourth order scheme is

$$\frac{\Delta t}{h} \leq \tfrac{3}{4}(c_p^2 + c_s^2)^{-\frac{1}{2}} \tag{3.321}$$

which is rather more stringent than the condition for the second order scheme.

As a guide to the accuracy of the fourth order scheme, for a compression wave travelling parallel to the x axis the ratio of the numerical wave phase velocity to the true compression wave velocity is

$$\begin{aligned}\frac{v}{c_p} = \frac{N}{\pi}\frac{h}{c_p \Delta t}\sin^{-1}\left\{\frac{5}{8}\left(\frac{c_p \Delta t}{h}\right)^2 - \right. \\ \left. -\frac{2}{3}\left(\frac{c_p \Delta t}{h}\right)^2 \cos\left(\frac{2\pi}{N}\right) + \frac{1}{24}\left(\frac{c_p \Delta t}{h}\right)^2 \cos\left(\frac{4\pi}{N}\right)\right\}^{\frac{1}{2}}.\end{aligned} \tag{3.322}$$

If we require one per cent accuracy in the phase velocity, we require about six points per wavelength in the fourth order method, compared with nine points per wavelength in the second order scheme, for a value of $\frac{c_p \Delta t}{h} = 0.6$. Working closer to the limit of stability requires rather more points per wavelength in both cases. The accuracy requirement means, however, that the increase in the time step Δt in passing to the fourth order scheme is not as large as was at first expected.

The increased complexity of the calculation at each point reduces the advantage of the fourth order scheme by a factor of about two. This figure will depend on computational details, but means that overall the saving in time for a given accuracy on passing from second to fourth order finite differences is about 50 per cent. This approach deserves further development: in particular, suitable boundary conditions need to be formulated. In all the examples described in this monograph the second order scheme has been used.

3.5.8 Finite difference calculations for large systems

In most practical cases, the size of the computer available dictates that the region that can be modelled by the finite difference method is smaller than the real system. This means that, in addition to the real boundaries of the system, the finite difference model will include artificial boundaries, and these will give rise to spurious reflections.

There are several approaches to this problem. The simplest is to ensure that the model is large enough for the spurious reflections to be distinguished, on the basis of their arrival times, from the reflections from real boundaries. This is practicable in some geometries, such as plates, but becomes very expensive for most cases. In many cases it is possible to remove some unwanted reflections by comparing calculations with and without the scatterer being studied: in this way the reflections of the incident wave from the artificial boundaries may be removed. For single-frequency waves it may be possible (Alabi 1985) to transform the coordinates so as to map an infinite region onto a finite domain, with a further change of variable to ensure that the radiation condition is satisfied. In other cases the size of the problem may be reduced by invoking symmetry arguments: for example, a normal stress applied to a plane surface will produce a field which is symmetrical with respect to the centre of the stressed area. It is also possible to set up boundaries on which the stress field has been calculated analytically, but the number of cases where this is possible are limited.

More generally, it is desirable to devise boundary conditions which will give rise to no reflections. There have been many attempts to do this, all of which start from analytic formulations of the wave equation and aim to derive boundary conditions which are equivalent to a radiation condition – that is, outgoing waves only. Some of the schemes proposed are not practical: Smith (1974), for example, requires the superposition of 2^n separately calculated finite difference solutions to eliminate the reflections from n boundaries. Lysmer and Kuhlemeyer (1967) proposed a scheme which uses viscous damping. This works successfully over only a rather narrow range of wavelengths, and is therefore not suitable for computations on pulses. Lindman's (1975) method was based on a Green function technique, but involved an element of parameterization in order to create non-reflecting boundaries: it was not entirely satisfactory. Other schemes have been derived by Bayliss and Turkel (1980) and Kausel and Tassoulas (1981). Engquist and Majda (1977,1979) have produced accurate schemes based on pseudodifferential operators, which involves the use of Padé approximants. They also make an assumption of paraxial propagation, so that their boundary becomes more highly reflecting at increasingly oblique incidence. For scalar waves this is adequate, as such waves do not propagate far into the interior after reflection, but the mode conversion which occurs with elastic waves means that a more effective scheme is desirable. Scandrett, Kriegsmann and Achenbach (1986b) have produced similar boundary conditions in a less mathematically forbidding way, basing their analysis on the cylindrical nature of the field radiated from finite scatterers in the far field. The method which offers the best compromise between complexity and effectiveness for general problems seems to be that of Clayton and Engquist (1977).

3.5.9 Finite difference calculations in complex geometries

When the geometry of the system to be modelled is such that all the boundaries lie along coordinate lines of the (usually square) grid, it is straightforward to apply the boundary conditions as described above. Boundary conditions specific to some simple extensions to this geometry have been worked out and applied: for example Georgiou and Bond (1985) have reported studies of slots near 120° corners, simulating root cracks in welds.

Complicated shapes can be built up by generating grids for small regions, and patching them together, but for smoothly curved boundaries other approaches are necessary. In such cases, a distorted grid is required, though the distortion is usually chosen so that the grid is nearly orthogonal near the boundary, to make the application of the boundary conditions easier. This is not strictly necessary, as it is possible to formulate conditions involving derivatives normal and tangential to the boundary as mixtures of differences along grid lines in and emerging from the boundary, but the accuracy is reduced if the departure from orthogonality becomes too great.

There are two approaches to the generation of curvilinear grids. In the first the function (in the present case, the displacement) is approximated by an interpolation function, which matches the values at the grid points, and this function is differentiated. The resulting expressions become very complicated on non-uniform grids. The second approach is a mapping method, in which the derivative with respect to the cartesian coordinates is transformed to terms involving derivatives with respect to the curvilinear coordinates. Here the accuracy of the final finite difference representation depends on the smoothness of the transformed grid (as well as on the smoothness of the differentiated function). Formulae for both approaches may be found in Hyman (1982): Thompson (1982) surveys the general field of curvilinear grid generation. The topic is also covered extensively in the review by Thompson, Warsi and Mastin (1982).

It is particularly important to realize that there are contributions to the error of a finite difference formulation which arise from rates of change of the grid spacing. In a hyperbolic system of equations, such as the elastic wave equations, errors will be propagated from the points in the grid at which they occur to other points throughout the mesh. For this reason the automatic generation of grids which are smooth in some well-defined sense, as discussed by Brackbill (1982), is particularly useful. Applications of the method are given in Saltzman and Brackbill (1982).

One particular problem when highly distorted grids are used is the choice of time step. To ensure stability, one has two choices. One can continue to use an explicit time-stepping scheme, in which the displacement at each node at a future time can be calculated independently of the displacements of other nodes at that future time. This is the approach that we have described above. The stability criterion dictates that no disturbance may propagate beyond the nearest mesh point in one time step, and this may necessitate excessively short time steps. The other approach is to use an implicit time-stepping scheme, in which the displacements at all the nodes at a future time are calculated by solving a set of simultaneous linear equations. This allows longer time steps to be used, but becomes prohibitively expensive for large problems. Blake (1985) has recently proposed a hybrid method, in which a small implicit region containing a distorted grid to model a defect is surrounded by an explicit region with a conventional square grid. His method was formulated as a finite element problem, but the same techniques could be applied within the context of finite differences.

3.5.10 Accuracy of finite difference calculation

There have been relatively few comparisons of finite difference results with analytic solutions, yet they are clearly important to the credibility of the method. Munasinghe and Farnell (1973) were mainly concerned with surface waves and used much finer grids than the ones suggested here. Their boundary conditions differed from the ones described above, in that they took centred differences for derivatives both normal and tangential to boundaries, instead of partially one-sided differences. In this section we compare the results of the second order finite difference scheme with analytic method for the scattering of bulk waves from ribbon-like defects inside a body, remote from any boundaries. The analytic results are those of Harumi (1961, 1962) who used an expansion in Mathieu functions: as explained earlier in the present chapter, these are the natural choice for this problem, which is a degenerate case of an elliptic cylinder, the eccentricity of the ellipse being infinite.

The cases for which the finite difference calculations were made were for plane compression waves incident on the face of a ribbon $2\lambda/\pi$ wide with a stress-free surface, on the edge of a stress-free ribbon $8\lambda/\pi$ wide, and on the edge of a rigid ribbon $2\lambda/\pi$ wide. The calculations were made in two stages: first the incident wave was set up in a finite mesh (see the remarks above about calculations on large systems) with no scatterer present, and then the results were subtracted from those with the scatterer present. This removes both the effects of the artificial boundaries and of the incident wavefield, leaving only the scattered wavefield as calculated by Harumi.

Figure 43 compares the amplitudes of scattered compression and shear waves for the compression wave incident on the face of the ribbon-like cavity of width $8\lambda/\pi$, obtained using finite difference calculations and analytical methods. The agreement is obviously good, especially when it is noted that the finite difference results used detectors only three wavelengths away from the scatterer whereas the analytical results are for the far field, and that the pulse was two cycles of sine wave which gives a spread of frequencies instead of the single frequency of the analysis. Numerical comparisons of the total cross-sections for scattering (Harker 1985) show that the finite difference results are in good quantitative agreement with those from the analytical method.

Figure 43

Amplitudes of scattered compression (solid) and shear (broken) waves for a compression wave of wavelength λ incident on the face of a ribbon-like cavity of width $8\lambda/\pi$. Figure (a) is the result of finite difference calculation, (b) gives the analytic result.

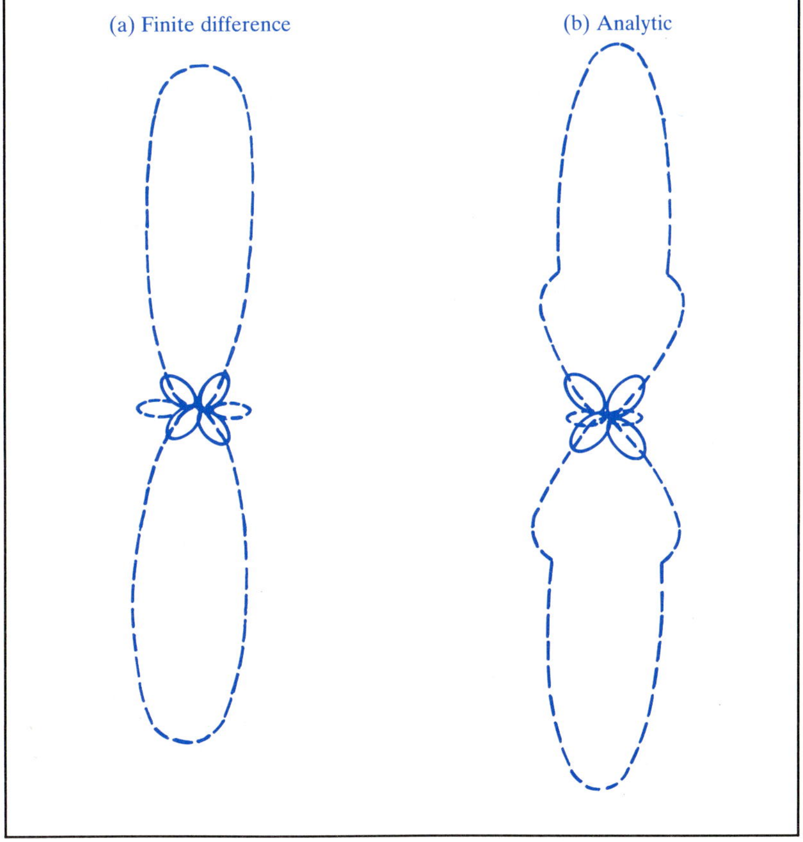

3.6 Comparative survey of methods

In determining the most appropriate computational method for a particular elastic wave scattering problem, there are several factors to bear in mind. In this section we concentrate on obtaining quantitative results: if physical understanding is the prime consideration, the emphasis will be different.

The first question to ask is whether the geometry is particularly simple, so that analytic results are available. In practice, the class of cases which it is *useful* to treat analytically is probably rather smaller than the class which *can* be so treated, when the complexity of the algebra and the difficulty of obtaining numerical values of some of the more obscure special functions are considered. For practical purposes, it is probably worth having plane-surface reflection coefficients, and the scattering from circular cylinders and spheres, ready to hand. Often it is the low frequency expansions of these exact analytic results that are of most use.

The next important question is whether the scatterer is small, weak, or small and weak. In those cases the Born approximation provides a simple way of calculating the scattering. The Born approximation and the quasistatic approximation are both low frequency theories, valid for values of ka less than unity.

If the frequency is high, that is, when the size of the defect is very large compared with the wavelength, the methods of ray theory are appropriate. Ray theory is most readily applied to crack-like defects, although the same principles can be applied to other geometries provided that the underlying canonical problems can be solved. The Kirchhoff theory is also applicable at high frequency, but is accurate only for a restricted range of angles of incidence. A combination of Kirchhoff and ray models, however, can handle most problems in the high frequency regime.

If none of the conditions above hold, one is inevitably faced with a lengthy numerical computation. This is the case in the intermediate region, where ka is of order unity. The best methods to use in this region are the t-matrix theories, finite difference or finite element methods. At the present state of the art, the t-matrix methods can probably handle a wider range of scatterer geometries than finite difference methods, but this is a matter of implementation rather than a difficulty in principle. Finite difference and finite element methods are particularly useful in cases where the scatterer is close to a surface: and this is one of the most common cases which nondestructive testing is required to study. Finite difference or finite element methods also have the advantage that they model short pulses easily, whereas the t-matrix theories are essentially single-frequency methods.

CHAPTER 4
Practical applications

4.1 Generation of elastic waves

The analytic results presented in Chapter 2 showed the main features of the generation of elastic waves in solids by surface and buried sources. The analysis is limited to sources of simple geometries, and to the limiting cases of step function, delta function or continuous wave sources. In practice transducers are only a few wavelengths across, and are excited by pulses of significant length, although the laser sources developed by Scruby *et al.* (1981) provide an elegant exception to this generalization. In most cases, then, the results have to be calculated by convolving the analytical results with pulse shapes and stress profiles. The finite difference method allows calculations with arbitrary sources and pulse shapes. As an example, consider the field of a strip source to which a tangential stress is applied.

Figure 44

The pattern of wavefronts arising from a surface source. A line source on the surface of a solid gives rise to cylindrical compression and shear wavefronts, with a plane front at the critical angle θ_0.

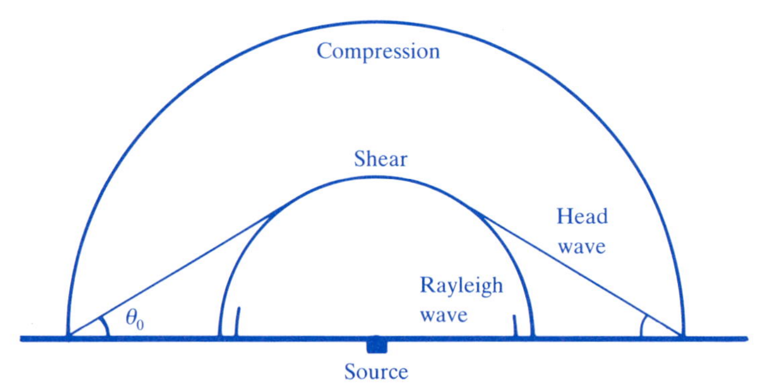

Figure 44 shows the wavefronts schematically. When this situation was modelled using the finite difference method, the source was a line source to which four cycles of a sinusoidal stress were applied, tangential to the surface. The resulting wavefield some time later is shown in Figure 45. In this plate, as in the other colour plates in this volume, the intensity of a wave at any point is represented by the intensity of the colour. Shear waves are represented in green, compression waves in red. Surface waves, which are composed of a mixture of compression and shear, show as orange or yellow. The edge of the region in which the calculation was carried out is shown by a blue line, and the region over which stress was applied to the surface (the transducer) is coloured turquoise. In Figure 45 the compression, shear, head and Rayleigh waves are all clearly visible.

Figure 45

The wavefield arising from a line source at the surface, to which four cycles of a sinusoidal stress are applied tangential to the surface. The intensity of a wave at any point is represented by the intensity of the colour. Shear waves are represented in green, compression waves in red. Surface waves, which are composed of a mixture of compression and shear, show as orange or yellow. The edge of the region in which the calculation was carried out is shown by a blue line. The compression, shear, head and Rayleigh waves are all clearly visible.

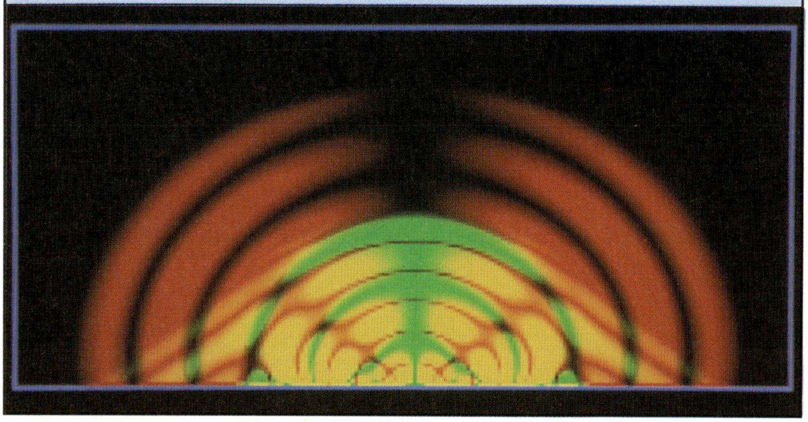

More quantitative results may be obtained from the time traces calculated in a finite difference calculation. Figure 46 shows the shear wave detected at various angles 3.3 wavelengths from the tangential surface source. The zero of angle is directly under the source. In Figure 46, for a material with wave speeds $c_p/c_s = 2$, the head wave is seen as a shear wave arrival before the cylindrical shear wavefront. The head wave only occurs above the critical angle θ_c, which is $\sin^{-1}(c_s/c_p)$. Figure 46 also shows that the amplitude of the cylindrical shear wave is much higher directly below the source than at high angles, in accordance with the calculations of Miller and Pursey (1954).

The angular variation of the compression and shear wave amplitudes is shown in Figure 47. The qualitative agreement with the analytic result of Miller and Pursey (1954) is good, with the exception of the peak at the critical angle.

The precise nature of the peak in shear wave amplitude at the critical angle deserves further consideration. Miller and Pursey (1954) perform their analysis in the far field, where all rays with a wavevector **k** which is parallel to the distance **r** from the source are seen at **r**. The large amplitude at θ_c could be interpreted as being due to the lumping of the whole of the head wave into the amplitude at that angle, instead of recognizing that it should be spread over the whole plane wavefront between the cylindrical compression and shear fronts. This appears to explain why Gupta and Kisslinger (1966) did not see the bump in the angular distribution experimentally, and would justify the approach

Figure 46

The displacements arising from a surface source. The source is a line source to which a four-cycle sinusoidal pulse of stress is applied, tangential to the surface. The figure shows the shear wave detected at various angles 6.7 shear wavelengths from the source. The zero of angle is directly under the source.

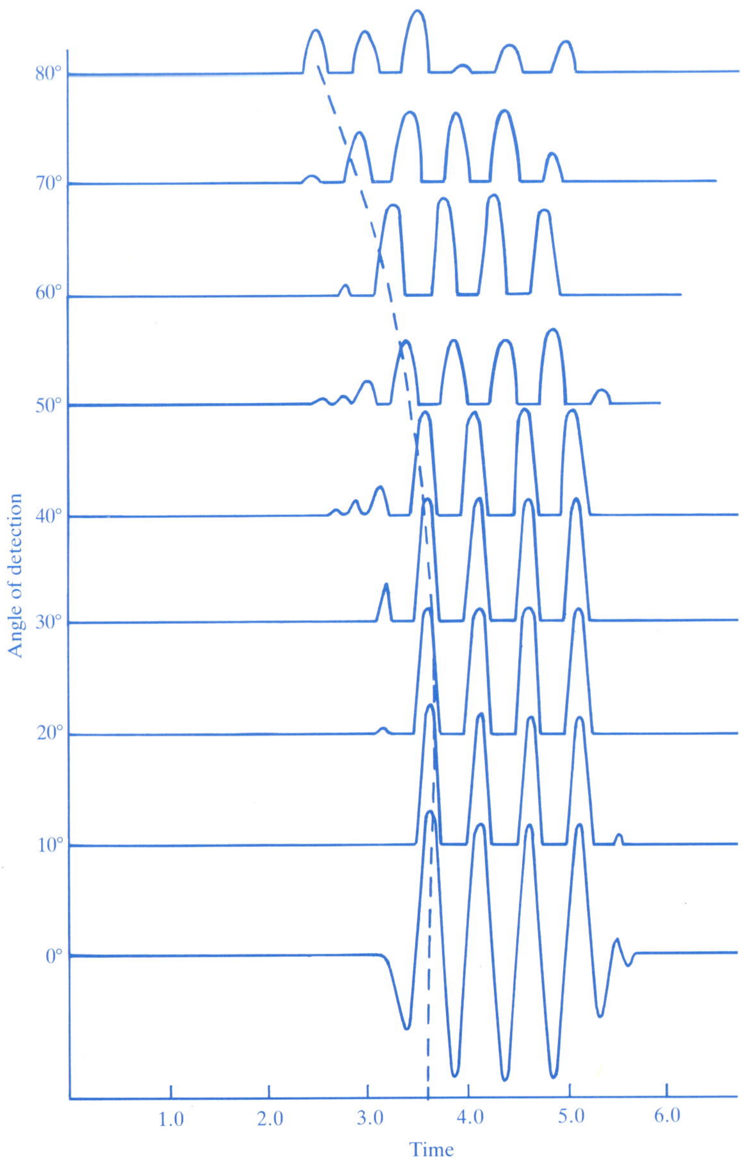

Figure 47

The angular variation of signals from a surface source. The source is a line source to which a four-cycle sinusoidal pulse of stress is applied, tangential to the surface. The figure shows the amplitudes detected as a function of angle 3.3 (upper part of figure) and 6.7 (lower) shear wavelengths from the source. The zero of angle is directly under the source.

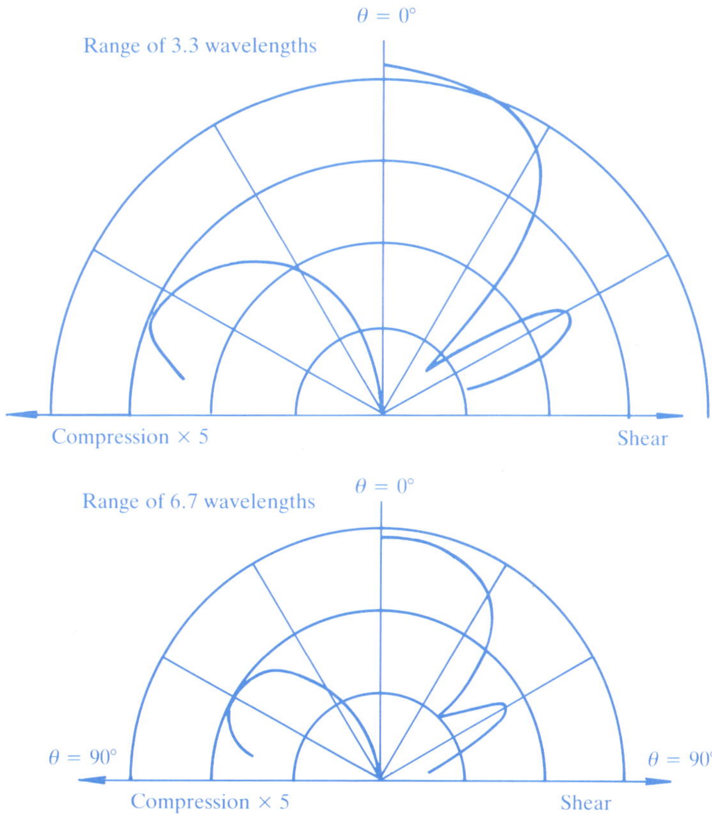

taken by Baborovsky, Slater and Marsh (1975) in their theoretical model. They used a modified Kirchhoff method, with an angular distribution which was that of Miller and Pursey (1954) but with the spike at θ_c removed.

4.2 Diffraction from a crack tip

The scattering from a semi-infinite crack was discussed in Chapter 3 as an example of a canonical problem in the context of ray theory. Crack tip diffraction has been used for some time for crack sizing, in the Time of Flight method (reviewed by Silk 1977). The method relies on the use of short pulses of ultrasound, so that the signals diffracted by edges of the crack may be resolved temporally. If the orientation of the crack is known, this allows its size to be calculated by using fairly simple geometrical arguments based on the ultrasonic path lengths.

Figure 48

The geometry of a Time of Flight diffraction experiment. Transducers are situated on opposite sides of a crack, to eliminate reflected signals and leave only diffraction.

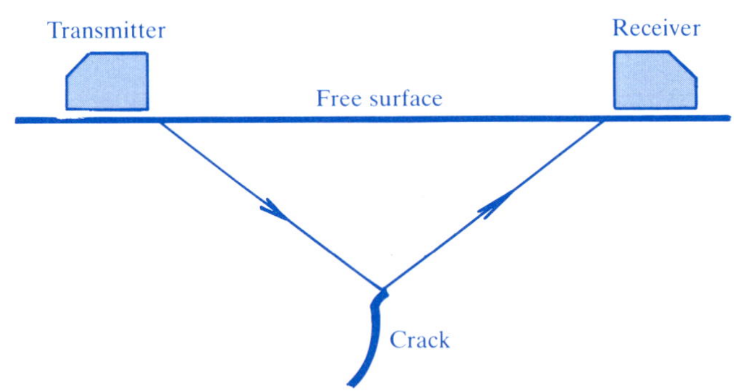

Figure 48 shows the geometry used in the Time of Flight technique: by arranging the transducers to straddle the crack and operating in pitch-and-catch mode, specular reflections from the crack faces are eliminated and only diffracted signals remain. The theory of the technique was treated by Ogilvy and Temple (1983) using an analytical calculation of Chapman (1981) (which gives the same results as Maue's (1953) theory). Earlier, however, the finite difference method was used for detailed comparison with experiments (Temple, Silk and Lidington 1979) and it is those results that we describe here.

Figure 49 shows the special polygonal test block that was constructed for the experimental phase of this exercise. A steel block was machined so as to provide several facets at different angles. A saw slit was cut almost to the centre, and a small slit was spark-eroded at the base of the cut to provide a narrow crack tip at the centre of the specimen. The experiments and the finite difference calculations used 2 MHz compression wave pulses. In all cases the transducers were placed symmetrically with respect to the crack tip, with the incident and diffracted directions both making an angle ψ with the normal to the slit.

The results of the calculation and the experiment are given in Figure

Figure 49

Diagram of a polygonal block constructed for measurements of the angular variation of diffraction from a crack, used to assess the time of flight diffraction method.

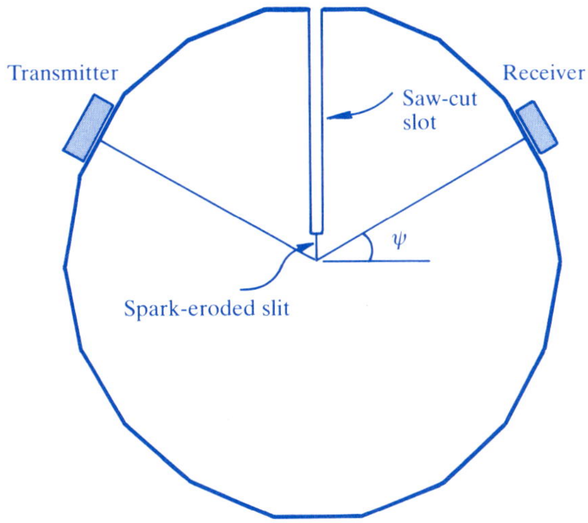

50: it will be seen that the agreement is good. The one exception is the minimum in the theoretical curve at about 40°, but the finite difference calculations agree with the analytical results (Ogilvy and Temple 1983) on this point. Ogilvy and Temple recommend that earlier calculations by Golan *et al.* (1980) should be treated with caution. Theory predicts, then, that there are distinct optimal angles for crack sizing with the Time of Flight technique. As multiprobe arrays of transmitters and receivers are often used with this method, it is generally possible to take advantage of these angles to increase the reliability of the inspection.

4.3 Scattering from a surface-breaking crack

From the point of view of structural integrity, one of the most important defects is a crack which breaks the surface. Such flaws can grow by several means, including fatigue, corrosion fatigue and stress corrosion, and their detection and sizing is central to many nondestructive testing processes.

Analytical methods are difficult to apply in the presence of a surface close to or intersecting the scatterer, and numerical methods probably offer the best general means of calculating the scattering in such situations. Recent calculations have used the t-matrix method (Bostrom and Kristensson 1980; Varadan and Varadan 1982; Varadan, Pillai and Varadan 1983) or numerical solution of singular integral

Figure 50

Amplitudes in the time of flight diffraction tests, as functions of angle. The circles with error bars show the experimental results, the smooth curve is the result of the finite difference calculation. The angle ψ is with respect to the normal to the face of the slit.

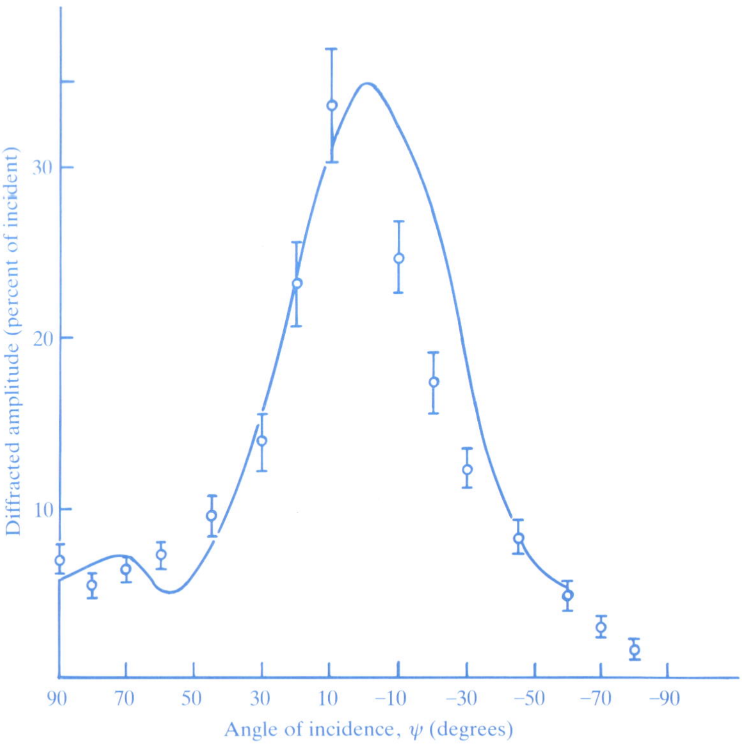

equations (Angel and Achenbach 1984; Achenbach, Angel and Lin 1984). Visscher's (1984, 1985) method is also based on the integral equation, but proceeds by expanding the crack-opening displacement in terms of a set of Gaussian basis functions. This jump in displacement is also taken as the fundamental quantity in van der Hijden and Neerhoff's (1984) treatment of general subsurface cracks. Our approach is through the finite difference method.

Before extracting quantitative information about the scattering, let us consider the effects which have to be included in a calculation. Here it is useful to use experimental visualizations to illustrate the processes. Several groups have used schlieren or photoelastic methods to obtain snapshots of the wavefield in elastic wave generation (Hall 1977) or scattering (Baborovsky, Marsh and Slater 1973a,b; Baborovsky, Slater and Marsh 1975; Hall 1976). Some progress is being made towards making such techniques quantitative, and eventually there may be close links between visualization and computation as is the coming practice in stress analysis (Sanford and Beanbein 1977), but for our purposes the qualitative results will suffice.

Figure 51

Interpretation of a photoelastic study of the scattering by a slit. The figure shows the wavefronts formed as a compression wave pulse is generated at the surface of a block and is scattered by a surface breaking crack. A detailed description is given in the text.

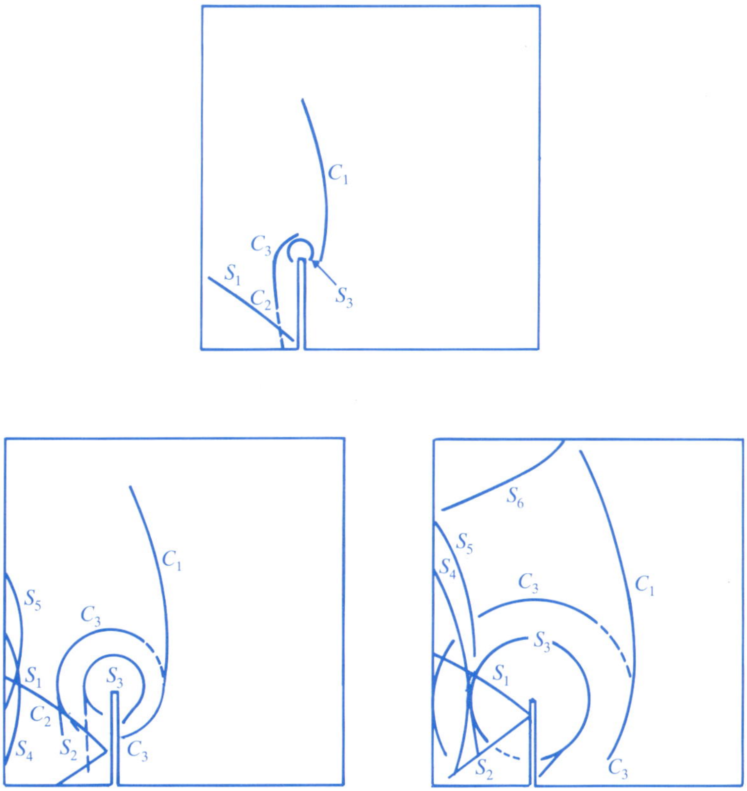

Hall (1976) used a photoelastic method to study the scattering of compression waves from a surface breaking slit. He interpreted his results in terms of wavefronts, and Figure 51 shows this interpretation for three different times after the injection of the compression wave pulse C_1. Part of C_1 is transmitted past the tip of the crack, part forms the reflected front C_2. The reflected front is continuous with the diffracted compression wave C_3, and a diffracted shear wave S_3 is also formed. As the incident compression wave interacts with the surfaces of

Figure 52

The compression (left) and shear (right) waves formed as a compression wave pulse is scattered by a surface-breaking crack, as calculated using the finite difference method. A detailed description is given in the text: compare Figure 51 and Figures 53 to 55

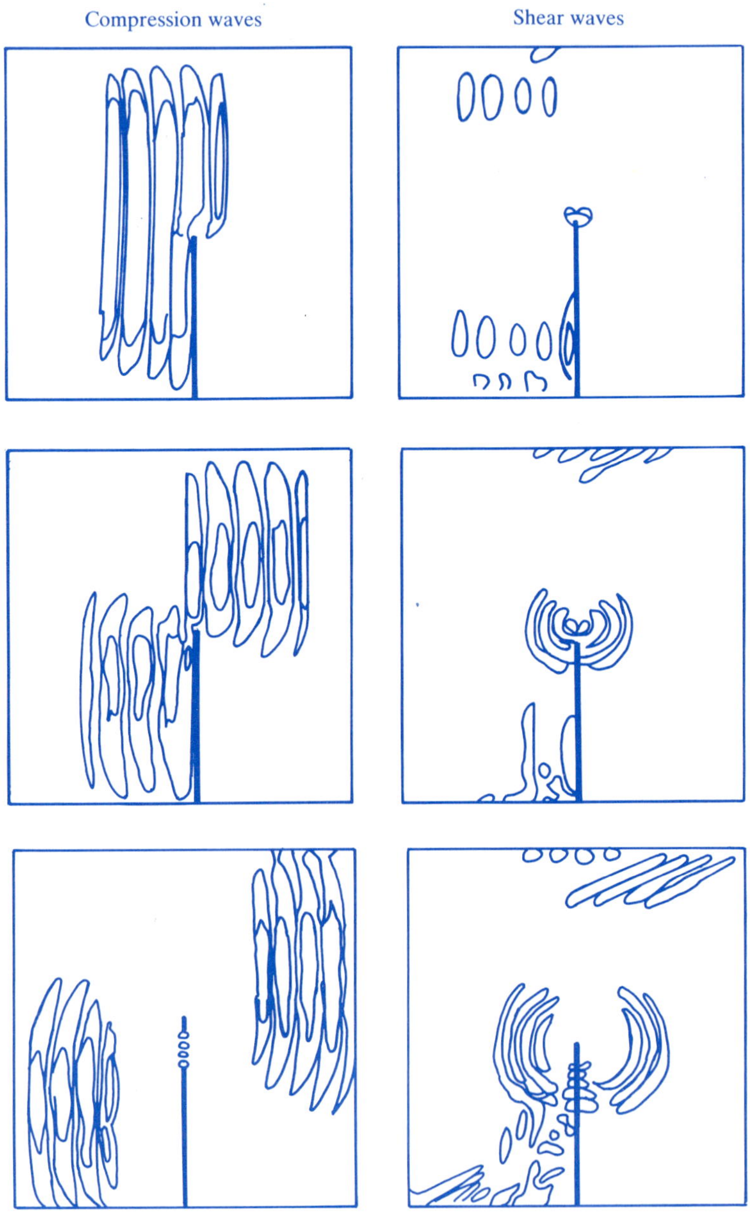

the block head waves S_1 and S_6 are formed, and a similar wave S_2 is formed by the reflected compression wave. The two remaining shear waves, S_4 and S_5 were formed at the edges of the transducer which generated the main compression wave C_1: this transducer was beyond the frame of the diagrams in Figure 51, to the left.

Figure 53

A compression wave pulse incident on a surface breaking crack, as calculated using the finite difference method. A detailed description is given in the text: compare Figures 51 and 52.

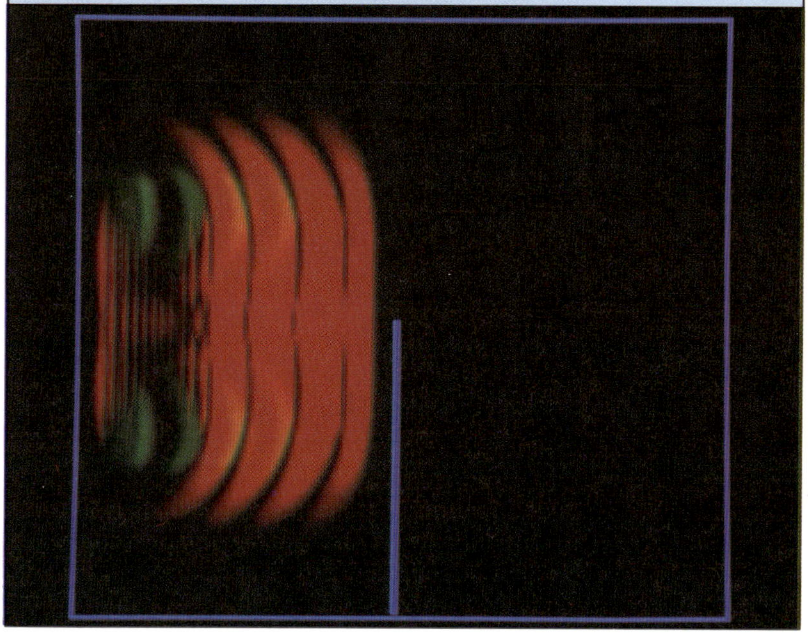

Figure 52 shows the results of a finite difference calculation of a similar scattering situation. Contours of constant amplitude are plotted, for the compression and shear components of the wavefield. The wavefronts shown in Figure 51 are clearly visible: the only exceptions are S_4 and S_5, which are absent because the compression wave in the model was set up inside the test block rather than being generated by a surface source. An even more graphic demonstration of the scattering is given in Figures 53 to 55. There, as in the other colour pictures in this volume, the intensity of a wave at any point is represented by the intensity of the colour. Shear waves are represented in green, compression waves in red. Surface waves, which are composed of a mixture of compression and shear, show as orange or yellow. The edge of the region in which the calculation was carried out is shown by a blue line. Clearly the qualitative agreement is excellent. The pedagogical value of animations of computer models such as these is just beginning to be exploited: films have been produced by several groups.

Figure 54

The compression wave pulse at a later time, when it has been scattered by the surface breaking crack. The fields were calculated using the finite difference method.

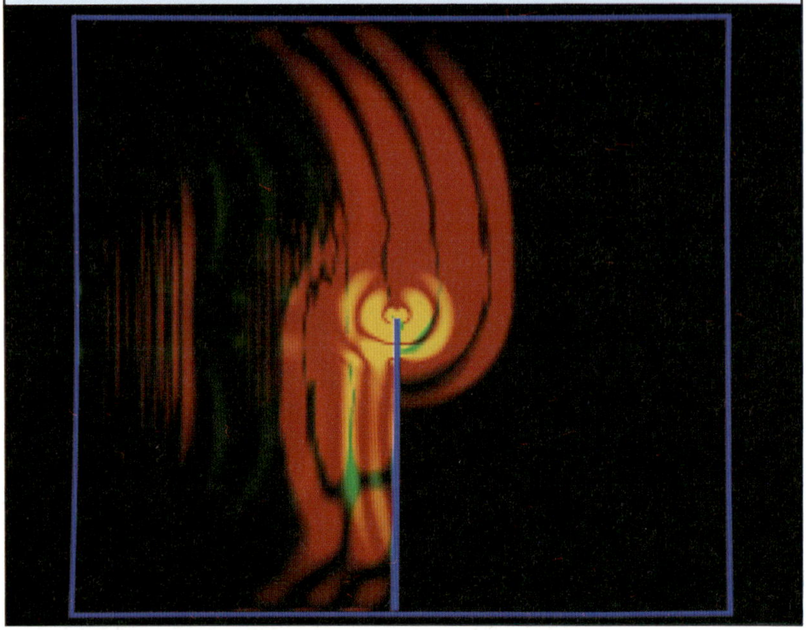

One feature of transducer behaviour is illustrated by this example. An attempt to set up a plane compression wave of finite width has resulted in small amounts of shear wave at the edges of the beam. This has been noted in many visualization studies (Hall 1977; Hayman 1977; Sachse, Hsu and Eitzen 1978; Weight 1982; Ying, Li and Zhang 1981). These are edge waves, which play an important part in any accurate model of transducer characteristics (Weight 1987).

Having established that the finite difference model does correctly model all the features seen experimentally in scattering from surface-breaking defects, and with knowledge of its accuracy as demonstrated in Chapter 3, we can with confidence extract numerical results from the method. Several cases have been studied, notably slots by Ilan, Bond and Spivack (1979) and Ilan and Bond (1979), and narrow cracks by Harker (1985) and Scandrett and Achenbach (1987). One of the motivations for the work of Harker (1984) was the review by Lumb (1977) of methods for testing pipelines. There it was concluded that elastic waves showed the greatest promise, but that the mode, angle and frequency had to be carefully chosen. As will be described further in Chapter 5, it is possible to distinguish between compression and shear waves on the basis of the arrival time of the signals, so either mode could be used. The choice of angle may further increase selectivity by reducing the amount of the unwanted mode that is

Figure 55

The compression wave pulse at still later time. Note that surface and head waves are clearly visible. The fields were calculated using the finite difference method.

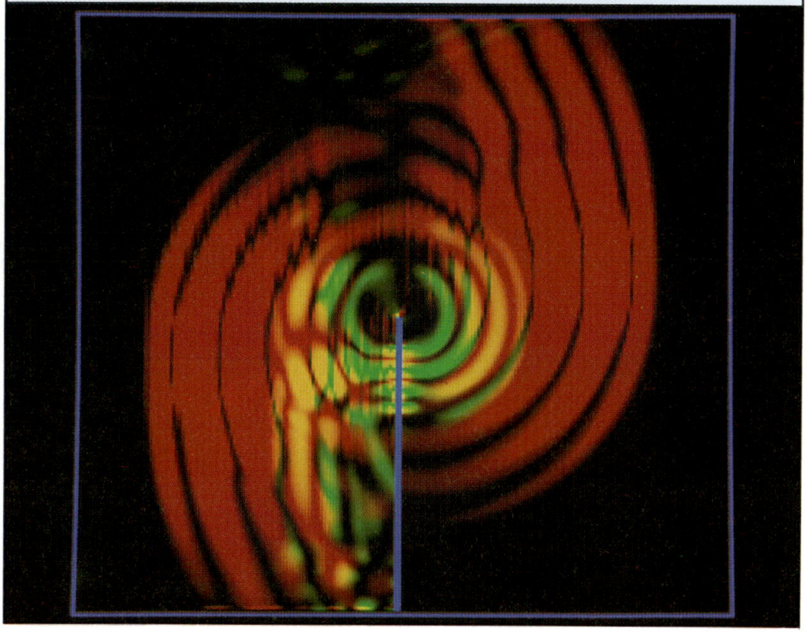

generated and reflected from the important defects. The choice of angle is complicated by diffraction, head wave and surface wave effects. It is not easy to make the choice of parameters on the basis of experimental work: the reviews by Lumb (1977) and Lumb, Bosselaar and Baborovsky (1978) compared the experimental results of Frielinghaus, Krautkrämer and Schlengermann (1970), Werneyer and Schlengermann (1971), Lidington, Saunderson and Silk (1974), and Baborovsky, Slater and Marsh (1975). All these groups agree that when shear waves at angles of 45° and above are used there is a non-monotonic relationship between the reflected signal and the crack depth. They differ, however, in their estimates of the magnitudes of the dips in the reflections coefficients: the work of Baborovsky, Slater and Marsh strongly suggests that this arises from difficulties in machining test slots or measuring natural flaws. A theoretical model removes these uncertainties and gives a true picture of the depth dependence of the reflection coefficient.

The calculations by Harker (1985) used a finite difference method to calculate the scattering of shear waves at angles between 45° and 75° from cracks up to 3 shear wavelengths deep: this was the range of most of the experimental results. Figure 56 shows the scattered wavefield for a crack 1.5 wavelengths deep, illuminated by a shear wave at 45°. Similar calculations for a range of crack depths show that there is a

Figure 56

The compression and shear waves formed as a shear wave pulse is scattered from a surface breaking crack 1.5 wavelengths deep, as calculated using the finite difference method. The wave was inserted by a transducer applied to the top surface.

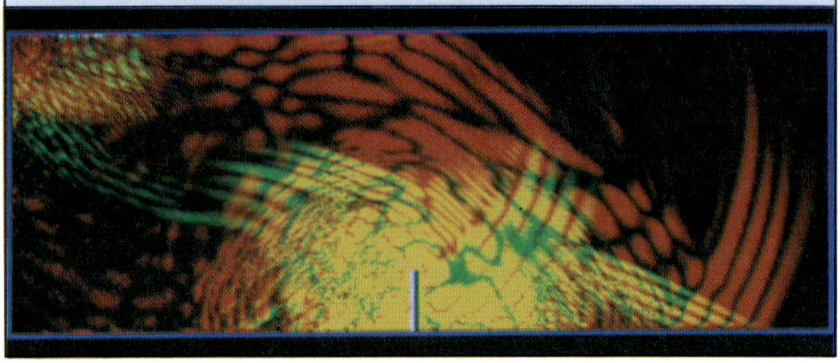

change from a large angular spread of scattered waves to a directed reflected beam as the crack depth increases. All modes, including surface and head waves, are generated by the scattering. The amplitude of the surface waves is quite insensitive to the depth of the crack and to the angle of the incident shear wave, but in the pulse echo geometry being modelled here the surface wave cannot return to the transducer. There is also a strong shear wave scattered from the crack and travelling normal to the surface of the plate: in thin plates this could give a long ring-down in the reflected signal.

With very small cracks, the scattered field is almost uniform in angle, arising from diffraction at the crack tip, but there is superposed on that isotropic field a directed field caused by the head wave. As a result of this, the direction of maximum shear wave scattering may not be the specular direction for small defects, but the angle of the head wave at the critical angle θ_c. As the crack depth increases, the specular character is established.

Figure 57 shows the amplitude of the shear wave reflected from surface breaking cracks as a function of crack depth. Curves labelled $\theta - \theta'$ show the reflection at angle θ' of an incident wave at angle θ. The results given in Chapter 2 for the reflection of plane waves as a function of angle would lead one to expect low reflectivity for 60° waves, as they are incident on the face of the crack at close to the minimum near the critical angle. This was indeed the case, and the signal at 45° was stronger: this is included in Figure 57. The results show that at a high angle (75°) the reflectivity is a monotonic function of crack depth, and the plateau in the 45° curve and the minimum in the 60° curve at a depth close to one wavelength agree with the conclusions of Lumb, Bosselaar and Baborovsky (1978). The quantitative behaviour, as measured by the change in reflected intensity between a $\frac{1}{2}$-wavelength and a 2-wavelengths deep crack, is in agreement with the experimental results, which showed a large range of values.

Figure 57

The amplitude (in arbitrary units) of the shear wave reflected from surface breaking cracks: curves labelled $\theta - \theta'$ show the reflection at angle θ' of an incident wave at angle θ. These results were calculated using the finite difference method.

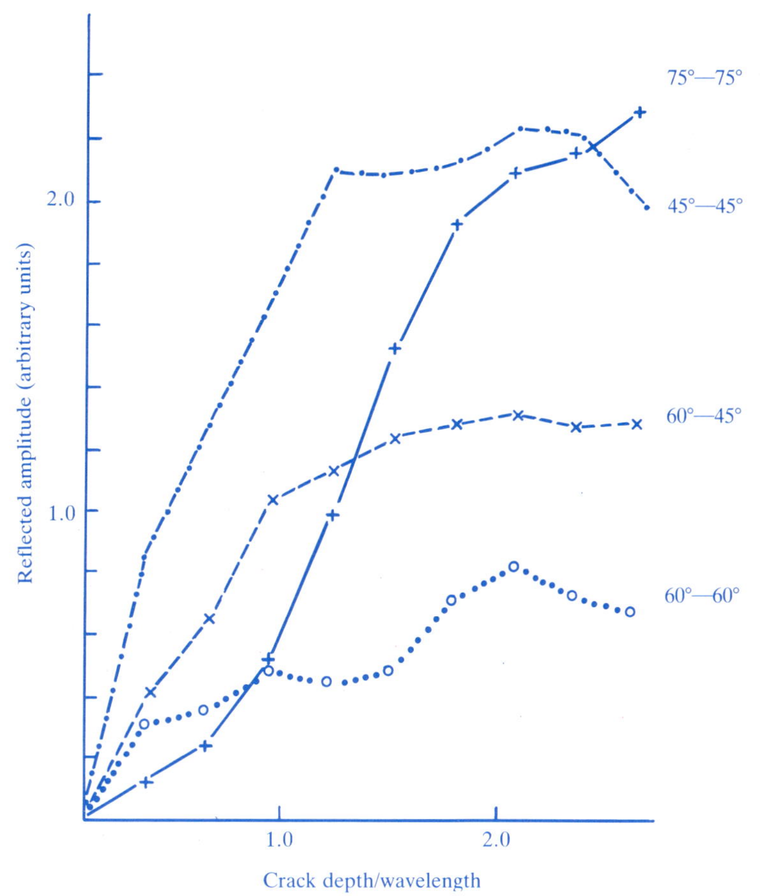

4.4 Scattering from a cylindrical cavity

In practical nondestructive testing it is often important to distinguish between the signals reflected by metallurgical features which do not detract from the mechanical properties of the material and stress-concentrating defects such as cracks. During the manufacture of a structure, slag inclusions or voids may form where welds are made, yet they may not affect the strength of the structure to any great extent. Such features are three dimensional, and may be quite smooth, in contrast to typical cracks which are two dimensional and have clearly defined edges. To distinguish slag and voids from cracks, then, it is necessary to understand the scattering from volumetric defects.

The basic ideas are fairly simple, and the Born approximation described in Chapter 3 gives arrival times for pulses which might be used to size volumetric defects. The Born approach does not, however, deal with the important class of creeping waves, which are bulk waves travelling around the interface between the defect and the matrix. The existence of creeping waves has been recognized for a long time: they have been proposed as a means of sizing objects at sea using sonar (Hickling 1962). The analytical theory of creeping waves was described in Chapter 3, and the present section gives a graphic demonstration of their existence, based on finite difference calculations.

The example we treat in this section is the scattering of elastic waves by a cavity in an isotropic solid. We choose wavelengths of the same order as the cavity radius, so that neither the quasistatic approximation nor geometric ray theory is applicable. We resort to an expansion method, as described in Chapter 3, although the t-matrix approach could also be used. To make the analytic approach simpler, we choose a cylindrical cavity.

Figure 58 shows the scattered wavefield calculated by the finite difference method. Figure 59 shows the same situation: the use of colour to represent wave mode is as before. In this example the cylinder had a radius $2.8/(2\pi)$ times the wavelength of the compression wave, and the incident wave consisted of a single cycle of sine wave. The elastic constants of the medium were such that the compression wave speed was twice the shear wave speed. In the finite difference calculation, the mesh spacing was one tenth of the wavelength, and the time increment was chosen so as just to satisfy the stability criterion. The radius of the cylinder was therefore 1.4 mm. The creeping waves are clearly visible as waves of large amplitude hugging the surface of the cylinder.

Figure 60 shows a similar calculation for a larger cylinder, with a radius of $8.4/(2\pi)$ compression wavelengths. The general features are very similar to those observed with the smaller cylinder. Figure 61 shows the same scattering process in colour.

It is a complicated matter to calculate the amplitudes of creeping waves analytically, especially for short pulses. Accordingly we shall not compare pulse shapes for analytical and numerical methods. We shall however compare the numerical and analytical results for the total scattering amplitude, to show that the finite difference method is accurate for calculations of this sort. The analytical results have already been calculated, as described in Chapter 3. For the finite difference calculations the stresses may be obtained directly by numerical differentiation of the wavefield, and total cross-sections may be calculated by numerical quadrature. In the present work, the energy flux was evaluated at 10° intervals around the cylinder. The following table compares the exact scattering cross-sections calculated by White (1958) with the finite difference results and with various of the

Figure 58

The scattering of a compression wave from a small cylindrical cavity. The amplitudes of shear waves (left) and compression waves (right) are depicted by contours. The radius of the cylinder is $2.8/(2\pi)$ compression wavelengths, and this picture shows the field 2.5 periods after a single cycle pulse impinged on the cylinder.

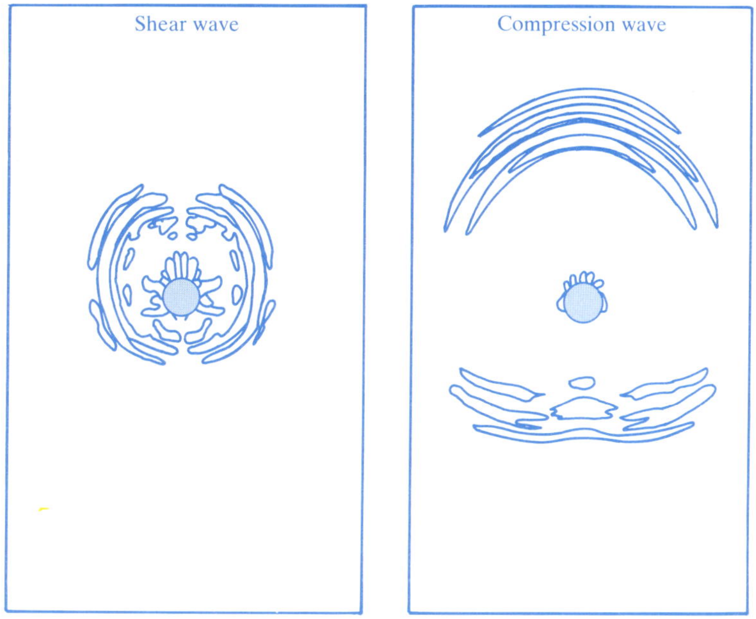

approximate methods described in Chapter 3. The cross-sections σ_{pp} and σ_{ps} describe the scattering of a compression wave into compression and shear waves respectively, whereas σ_p is the total $\sigma_{pp} + \sigma_{ps}$.

Scattering cross-sections of a cylinder for incident compression waves

Calculation	$\sigma_p/2a$	σ_{pp}/σ_{ps}
Exact	2.2	2.6
Finite difference	1.9	2.5

Although the finite difference results are calculated in the relatively near field of the cylinder (only about 2 radii away from the centre) and

Figure 59

The scattering of a compression wave from a small cylindrical cavity. The amplitudes of compression waves are shown in red, shear waves in green. The radius of the cylinder is $2.8/(2\pi)$ compression wavelengths, and this picture shows the scattered field 2.5 periods after a single cycle pulse impinged on the cylinder.

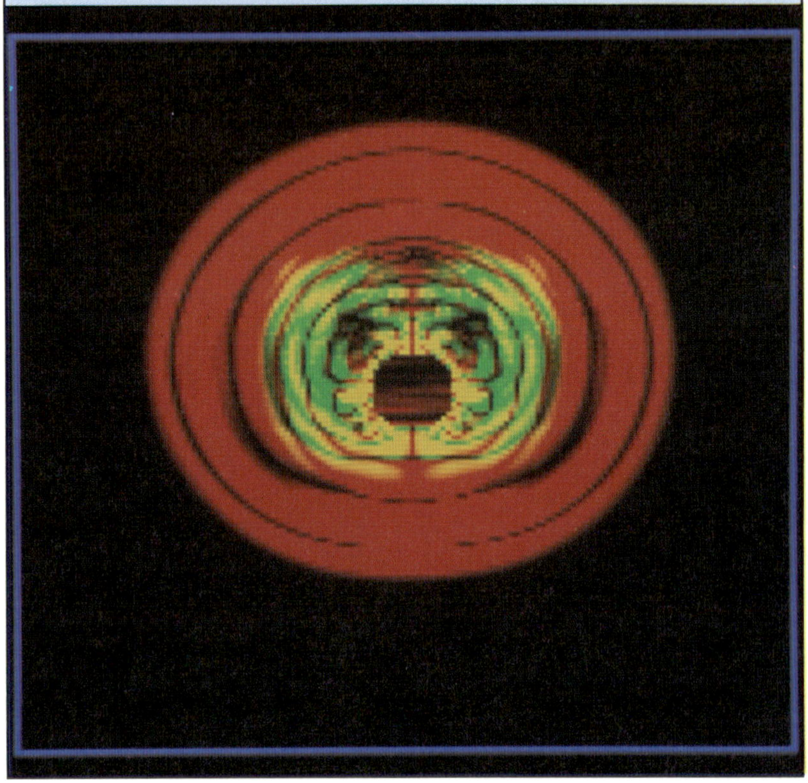

the finite difference pulse is short (only one sine wave cycle), giving quite a wide range of frequencies, the finite difference results are in very good quantitative agreement with the exact ones.

The details of the scattered waves can be extracted from plots of the displacements in the scattered waves as functions of time. Figure 62 shows the geometry of an arrangement in which the displacements of the waves scattered back along the input wave direction are monitored. If the radius of the cavity is a, the initial distance of the wavefront from the centre of the void is x and the distance from the centre of the void to the detector is d, the expected arrival times by various paths are as follows, for an incident compression wave:

direct compression wave

$$t_p = (x-d)/c_p; \tag{4.1}$$

Figure 60

The scattering of a compression wave from a large cylindrical cavity. The amplitudes of shear waves (left) and compression waves (right) are depicted by contours. The radius of the cylinder is $8.4/(2\pi)$ compression wavelengths, and this picture shows the field 2.5 periods after a single cycle pulse impinged on the cylinder.

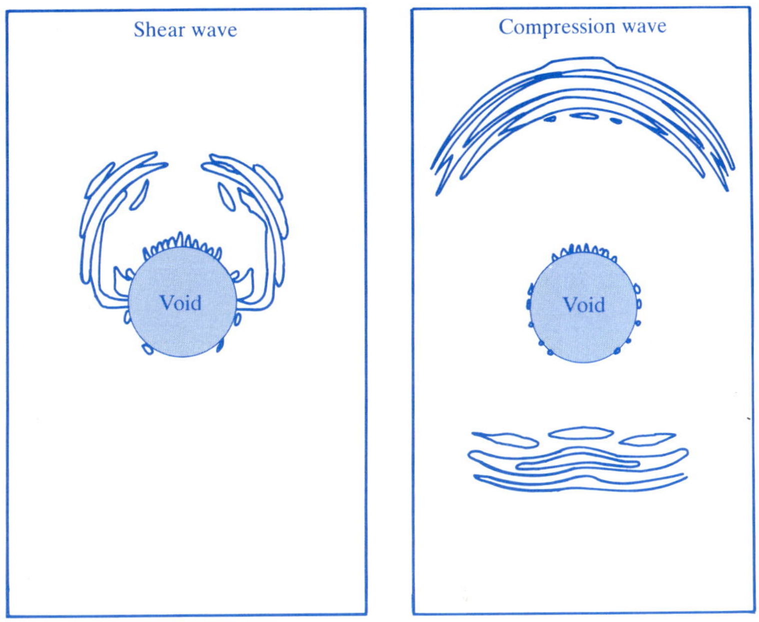

specular compression wave

$$t_{pp} = (x+d-2a)/c_p\,; \tag{4.2}$$

creeping compression wave

$$t_{pp'} = (x+d)/c_p + 2\pi a/c_p\,. \tag{4.3}$$

The last equation shows that the creeping wave is to be thought of as being generated where the incident compression wave is tangential to the cylinder, creeping around half the circumference, and emitted again tangentially. Figure 63 shows the computed time domain response of the cylindrical cavity to a short input pulse: the direct reflection and the first creeping wave arrival time are clearly visible. Analysis of pulse arrival times in this way has been suggested by Pao and Sachse (1974) as a means of determining the sizes and properties of inclusions in materials, and they give theoretical results which show promise.

Figure 61

The scattering of a compression wave from a large cylindrical cavity. The amplitudes of compression waves are shown in red, shear waves in green. The radius of the cylinder is $8.4/(2\pi)$ compression wavelengths, and this picture shows the scattered field 2.5 periods after a single cycle pulse impinged on the cylinder.

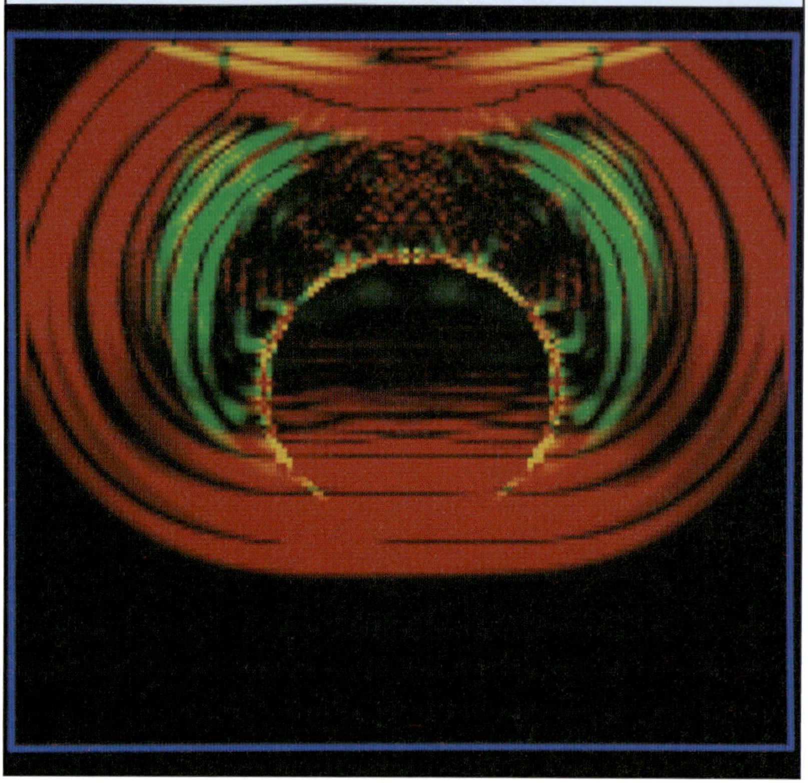

4.5 Scattering of Lamb waves by a surface-breaking crack

The dispersive nature of Lamb waves makes the calculation of their generation and propagation complicated except when they are detected far from the source and when the frequency is chosen so as to be close to a stationary point of the dispersion curve. Even in such cases, it is difficult to compute the scattering of Lamb waves from defects of practical importance. The reason is obvious: if the wavelengths of Lamb waves are comparable with the thickness of the plate in which they propagate, and defects become significant when they penetrate to a depth of the same order as the plate thickness, then the scattering problem lies in the most difficult region where the wavelength is comparable with the defect size.

Analytical methods can be applied to Lamb wave scattering but they become very complicated. For example, Merkulov and Rokhlin (1970) and Rokhlin (1980a,b) used modified Wiener–Hopf methods; Auld and

Figure 62

Geometry for time domain analysis of the scattering from a cylindrical cavity, showing the direction of the incident wave and the position at which the displacements are detected.

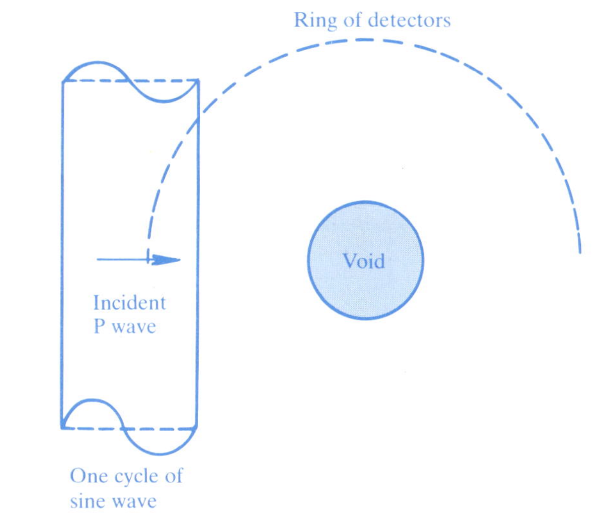

Tsao (1977), Auld and Tan (1977) and Tan and Auld (1980) applied a variational technique; and Rokhlin (1980a,b) has employed a multiple diffraction method. Auld and co-workers considered cracks normal to the plate surface, but Rokhlin treated defects parallel to the surface. Analytic solutions are not available for more general defects. Numerical methods are likely to be the most useful for Lamb wave scattering: Koshiba, Karakida and Suzuki (1984) have given results using a finite element method, and here we show how the finite difference technique may be applied. Georgiou and Bond (1985) have also presented results for Lamb wave scattering problems using finite differences.

As an example, we consider the scattering of Lamb waves in a steel plate by a surface-breaking crack. The generation of the waves is by a line source of shear stress. It is quite easy to apply shear stresses to steel plates using non-contacting electromagnetic transducers, by using coils to induce eddy currents in the steel in the presence of a strong magnetic field perpendicular to the plate. This is illustrated in Figure 64. In practical applications, greater wavelength selectivity can be achieved by using a flat coil with several turns, but the single line illustrates the principle. For more complicated generating coils, the finite difference method may be easier to apply to predict the waves generated than the analytic method of Shkarlet (1974).

Figure 63

The figure shows the compression wave scattered in the backward direction from a cylindrical void. The radius of the cylinder is $8.4/(2\pi)$ compression wavelengths. The inset shows the incident single-cycle pulse.

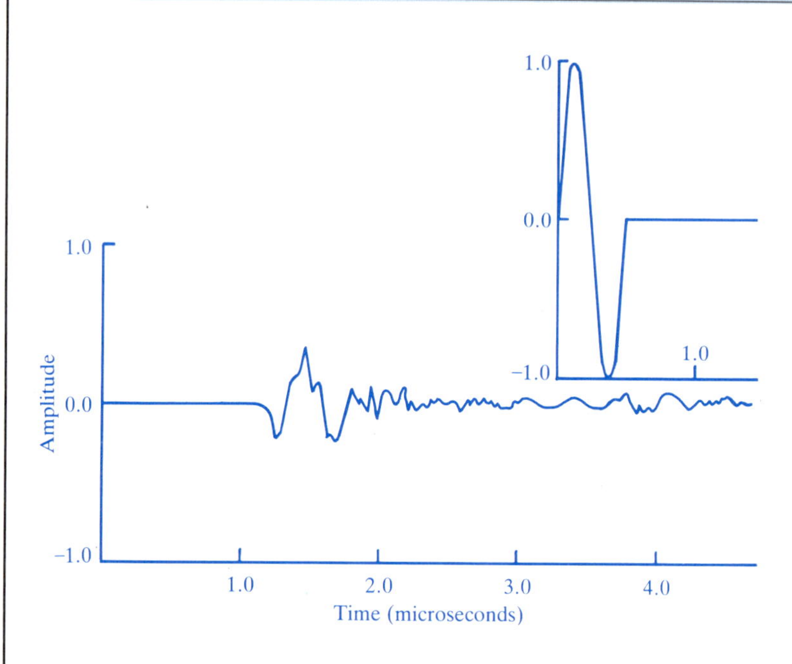

With the shear stress applied to a strip on the surface of an aluminium plate 12 mm thick, at a frequency of 0.375 MHz, the figures in Chapter 2 show that the dispersion curve for the s_1 mode is close to a maximum. The dimensionless parameter $\Omega = \omega t/2c_s$ has the value 4.25. In this case (see also Harker 1985) the wavelength and plate thickness are comparable, so the displacements are more readily displayed as a distorted grid than as separate compression and shear fields. Figure 65 shows the displacements in the plate 2.5 μs after a pulse of frequency 0.375 MHz was applied to the source shown. The stress at the source was tangential to the surface of the plate.

It is immediately apparent that the wave which is generated is symmetrical about the source, and is a superposition of symmetric and antisymmetric modes. The input pulse spreads, but after quite a short distance (50 mm for the detector shown in Figure 65) the s_1 mode is readily identifiable, as shown by Figure 66.

For calculations of scattering it is useful to be able to separate the symmetrical and antisymmetrical modes. This is easily done with a finite difference model by using pairs of sources and detectors on opposite sides of the plate. Using the same frequency of 0.375 MHz,

Figure 64

The electromagnetic generation of shear stress. A conductor carrying an alternating current i produces a magnetic flux density $\mathbf{B}'$, which induces a current density $\mathbf{j}$ in a conducting plate. With a strong magnetic flux $\mathbf{B}_0$ applied normal to the plate, there is a volume force $\mathbf{f} = \mathbf{j}_\wedge \mathbf{B}_0$.

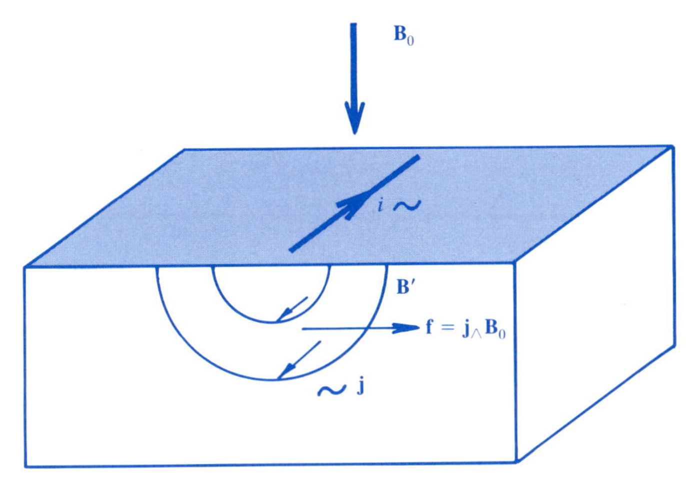

Figure 65

The displacements in an aluminium plate 12 mm thick 2.5 μs after a pulse of frequency 0.375 MHz was applied to the source shown. The stress at the source was tangential to the surface of the plate.

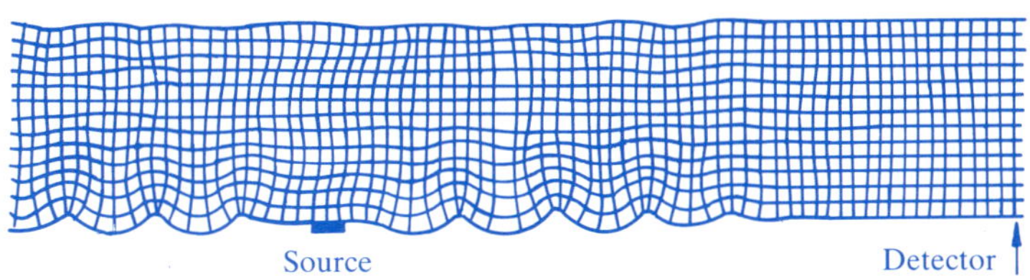

the reflection and transmission of a_1 and s_1 modes, and the mode conversion between them, was calculated for several crack depths. Figure 67 shows the results for incident a_1, and Figure 68 is for incident s_1. In these figures, we use T to denote transmission past the crack and R for reflection from the crack. Then, for example, a_1Rs_1 denotes the amplitude of the s_1 mode reflected from a crack when an a_1 mode is incident.

From the results in Figures 67 and 68 it is possible to draw some

Figure 66

The displacements parallel to the surface at the position of the detector of Figure 65, which was 50 mm distant from the Lamb wave source. The s_1 mode gives a pulse which can be clearly identified.

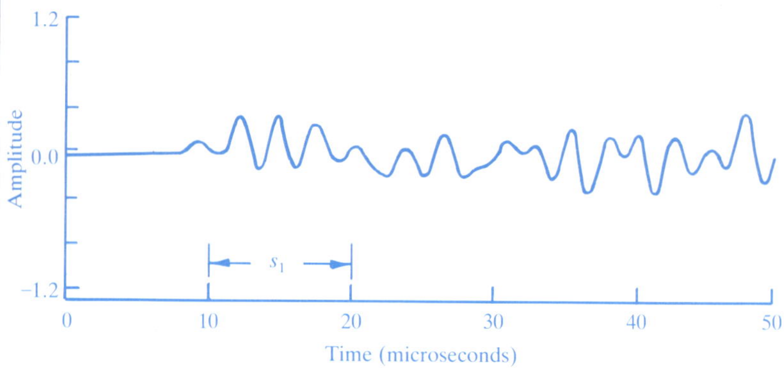

conclusions about the usefulness of these Lamb waves for nondestructive testing applications. It is clear that they are very sensitive to surface-breaking cracks, and therefore have potential for detecting defects. Unfortunately the curves rapidly saturate as functions of depth, so these modes would not be useful for determining the depths of cracks. With the modern emphasis on defect characterization as opposed to detection, these modes could not be recommended. Other modes do offer potential for depth determinations: s_2 is particularly attractive. As the mode index increases, however, so does the number of modes that can be excited at a particular wavelength. This makes unambiguous detection of reflections of higher modes difficult. Another problem is that, because the Lamb wavelengths are quite long, accurate positional information about defects is hard to obtain. Further, one cannot normally distinguish between defects on the same side of the plate as the source and those on the opposite side.

4.6 Problems in engineering materials

The problems of the propagation, reflection and scattering of elastic waves that we have considered so far have all been somewhat idealized, in that we have ignored the detailed structure of the materials of which metallic structures are made. This may have significant effects on the detectability of defects.

There are two effects that we should consider. First, the material is not homogeneous and isotropic, but is formed of a large number of small grains, each of which is not isotropic but shows the symmetry of the crystal structure. Secondly, the defects in the material may not present smooth faces: stress corrosion cracking, for example, follows

Figure 67

The amplitudes of a_1 and s_1 Lamb waves reflected from and transmitted past a surface-breaking crack when an a_1 mode is incident. The Lamb waves are characterized by a dimensionless parameter $\Omega = \omega t/2c_s = 4.25$.

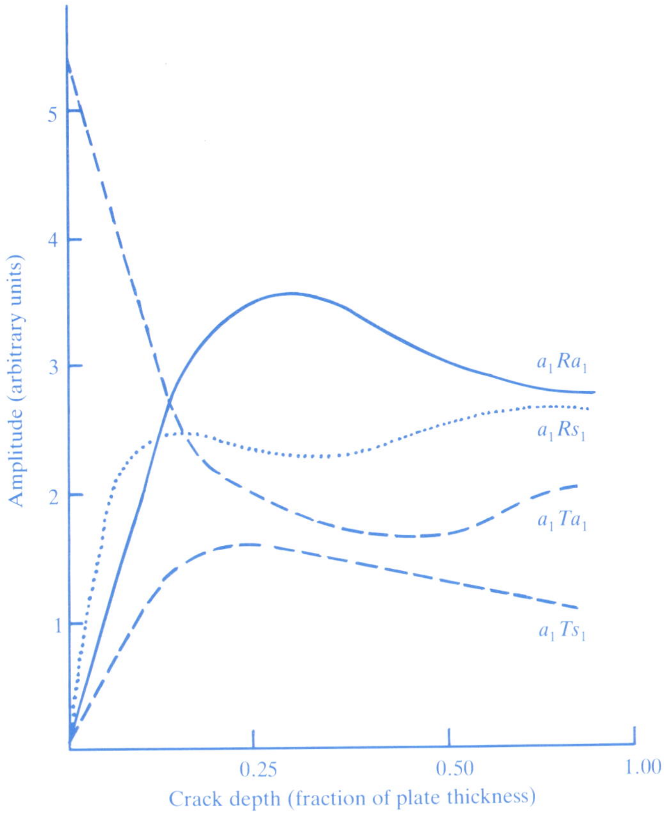

paths between the grains of the material, and the crack faces will therefore be rough on the scale of the grain size. A further complication is that the crack faces may not be completely separated: in the case of stress corrosion cracking there may be particles of corrosion product bridging the crack at some points, and in fatigue cracking relaxation of the stress may have caused the faces to come into contact. It is important to consider how these considerations may affect the results we have presented so far.

Figure 68

The amplitudes of a_1 and s_1 Lamb waves reflected from and transmitted past a surface-breaking crack when an s_1 mode is incident. The Lamb waves are characterized by a dimensionless parameter $\Omega = \omega t/2c_s = 4.25$.

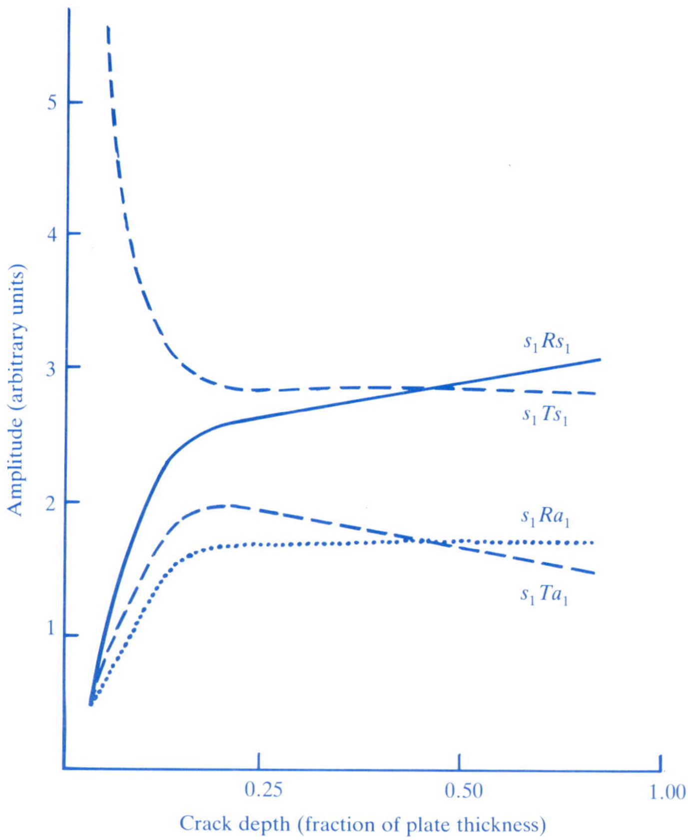

4.6.1 Attenuation and scattering in polycrystals

When calculating the properties of elastic waves in polycrystals considerable simplifications can be made if allowance is made for the fact that metallic structures tend to be made from quite finely grained materials, to increase the resistance to brittle crack propagation (Knott 1973). In the steels used in the United Kingdom gas arterial network, the grains are a few microns in diameter. In addition, one may note that the variations in density from grain to grain are small, as are the differences in elastic constants. A first order scattering theory is likely to be sufficiently accurate.

The first calculations of the attenuation of waves in granular media were made by Lifshitz and Parkomovsky (1950), and independently by Bhatia and Moore (1959). They define deviations of elastic constants from the averages over the entire crystal by

$$\delta c_{ijkl} \equiv c_{ijkl} - \langle c_{ijkl} \rangle, \tag{4.4}$$

and density deviations

$$\delta\rho \equiv \rho - \langle \rho \rangle. \tag{4.5}$$

The Born approximation of Chapter 3 may now be applied.

If we assume that the grains are sufficiently small that the shape factor $S(\Delta\mathbf{k})$ may be approximated by the volume of the grain, and that the density changes are negligible, then the scattering amplitude of an incident plane wave by a grain of roughly spherical shape and radius a is given by

$$f_i(\mathbf{k}^s) = -\frac{|k^s|^3}{3\rho\omega^2}\delta c_{ijkl} k_j^s k_k^i u_i^i a^3, \tag{4.6}$$

for incident wavevector $\mathbf{k}^i$ and scattered wavevector $\mathbf{k}^s$, and incident wave amplitude $\mathbf{u}^i$. Note that, as before, $\mathbf{f}$ is the coefficient of the dimensionless outgoing spherical wave. If the grains have random orientations we may evaluate the total energy scattered out of the incident beam by averaging the energy scattered by one grain over all possible grain orientations, and summing the result over the number of grains. It is clear that the result will depend on the fourth power of the frequency (from $(k^s k^s k^i k^s/\omega^2)^2$). The grain size dependence will vary as $\langle a^6 \rangle$ from the energy expression, and as $1/\langle a^3 \rangle$ from the number of grains. The terms which arise from the angular averaging of the elastic constants have been calculated for a variety of crystal systems by Papadakis (1965,1968). For cubic systems the attenuation of compression waves is given by

$$\alpha_p = \frac{(c_{1111} - c_{1122} - 2c_{1212})^2 \rho^2}{1125 Y^4}\left[2 + 3\left(\frac{c_p}{c_s}\right)^5\right]\frac{\langle a^6 \rangle}{\langle a^3 \rangle}\omega^4. \tag{4.7}$$

Here Y is the Young's modulus of the polycrystal, and the wave speeds c_p and c_s are also those appropriate to the average material.

Similar expressions have been obtained by Hudson (1968) and Hirsekorn (1982) using different approaches. Hudson considers the consequences of the fact that the transitions between grains are sharp, rather than smooth as the theory of Lifshitz and Parkomovsky assumed. He highlights the difficulties of defining appropriate correlation functions describing the way in which the elastic constants at one point in the material are related, in a statistical sense, to those at another point. Hirsekorn's treatment goes beyond the approximation of small grains. Hudson (1968) also proves the assertion made in Chapter 2 that the polycrystalline material may be treated as an effective isotropic medium. He gives formulae for the elastic constants λ and μ in terms of the crystalline elastic constants c_{ijkl}. Mason and McSkimin (1948) give similar formulae, but Hudson makes some corrections to their results. Evans *et al.* (1978) noted that the strong wavelength dependence of the scattering means that the attenuation is likely to be dominated by the largest grains, and developed an extreme-value theory to account for this. In metals, however, the grain size is likely to be more closely controlled than in the ceramics they were considering.

If the wavelength of the elastic waves is small compared with the grain size, the scattering process may be regarded as a large number of partial reflections from interfaces between grains. In such cases the scattering is independent of frequency (as in the plane surface reflection coefficients of Chapter 2), and the number of interfaces encountered by a wave on its way through a polycrystal is inversely proportional to the grain radius. Sometimes the grain size becomes so large, and the structure is so well characterized, that it is feasible to make detailed ray tracing calculations in the structure. The highly directional grain structure of welds in austenitic stainless steels is a good example (Ogilvy 1985).

For the intermediate region, when the wavelength is comparable with the grain size, Lifshitz and Parkomovsky (1950) give formulae for the attenuation which are proportional to the grain radius and to the square of the frequency.

Rokhlin (1972) derived a theory which has the correct functional form in all three wavelength regimes. Although Rokhlin's quantitative results were rather poor, his theory has been applied in an ingeneous way by Beltzer and Brauner (1985). They redefine some of the constants of Rokhlin's theory so as to obtain the correct limiting behaviour, and use the resulting attenuation as a function of frequency to calculate, through the Kramers–Kronig relations, the wave speed as a function of frequency.

Experimental data (Papadakis 1968, 1984) seem to bear out the theoretical predictions for materials which are on the average isotropic and homogeneous, with equiaxed single-crystal grains. It is clear that for small grains, less than 1 per cent of the wavelength in size, the attenuation is negligible. Reynolds and Smith (1984) have noted that the additional attenuation arising from dislocation damping should be subtracted from the measured total if microstructural effects are to be studied. Figure 69 shows how this may be done, and also demonstrates that the variation of attenuation with frequency depends not just on the average grain size but also on the distribution of grain sizes.

In most applications, the frequency of the elastic waves used will be chosen so that the attenuation is low: that is, so as to remain in the weak scattering regime. In some cases this is not possible (for example,

Figure 69

A plot of attenuation α divided by the square of the frequency w against the square of the frequency allows dislocation damping to be separated from grain boundary scattering in a polycrystalline material. Note that the shape of the curve is strongly dependent on the distribution of grain sizes, not just on the average grain size $<d>$.

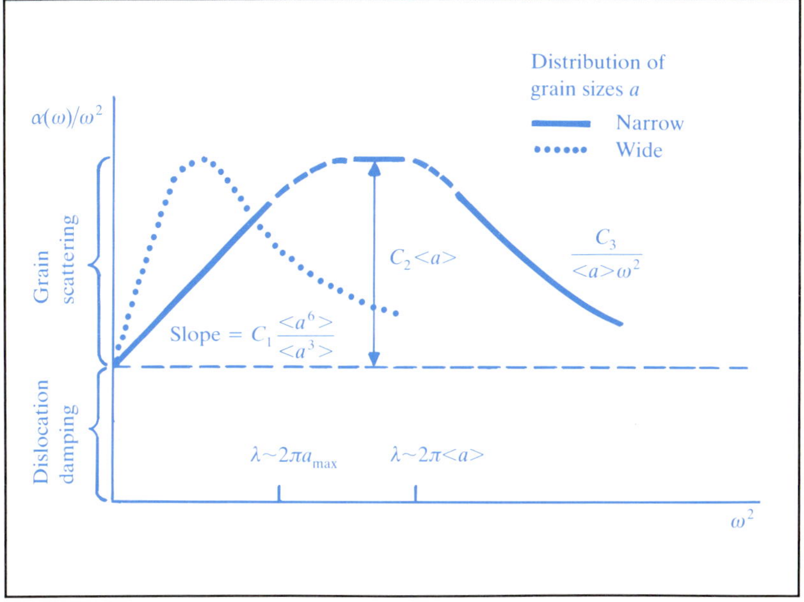

when composite materials are to be tested) and in other applications high attenuation is desirable (in backing materials for transducers, to reduce the ringing which distorts short pulses). It then becomes important to consider multiple scattering, that is, the effect of an elastic wave that is scattered by one grain and then again scattered by another. This problem was addressed, in the context of scalar waves, by Foldy (1945), Lax (1951, 1952), and Waterman and Truell (1961), among others. These workers, however, neglected the correlations between scatterers. Recently attempts have been made to include correlation effects, producing self-consistent theories (Datta 1977b; Varadan, Varadan and Pao 1978; Korringa 1979; Devaney 1980; Sayers 1980, 1981; Sayers and Smith 1982). Willis (1981) has developed a variational method which is designed to make the best use possible of the available correlation data: that is, a better approximation would require higher order correlations. This approach is clearly well suited to practical problems, where the information on the correlation functions may be quite limited. Weaver (1985) has also used a variational method.

Practical use may be made of the frequency variation of attenuation to probe the microstructure of materials. The most obvious application (Goebbels 1975, 1976; Goebbels and Höller 1976, 1978) is to the determination of grain size. It can be particularly useful to locate regions where the average grain size changes, as these may indicate changes in metallurgy. If a specimen is examined from different directions it is possible to deduce the texture of the material, and therefore to monitor the action of rolling mills or the rate of cooling of castings.

4.6.2 Reflection at rough surfaces

The surfaces of real materials will be rough on several scales: the atomic scale need not concern us here, but above that will be effects of machining of the surfaces, grain sizes, random paths through grains at fracture surfaces, and probably larger-scale undulations.

Mandelbrot, Passoja and Paullay (1984) have argued that a fracture surface will be a fractal surface, exhibiting the self-similar property of fractals (Mandelbrot 1977, 1982) down to the atomic scale. The formal properties of monochromatic waves scattered by fractals have been studied by Berry (1979), and Berry and Blackwell (1981) have treated pulses. The scattered waves are known as diffractals. In practice, the scattering of waves with a narrow frequency distribution from a rough surface will be affected by a fairly narrow range of scales of roughness, and the fractal nature of the surface may be ignored. Nevertheless, it is necessary to quantify the roughness in some way. The range of parameters used to characterize rough surfaces is enormous: Thomas (1982) gives a useful survey of the field. Greenwood (1984) has presented a uniform theory of roughness based on the assumption that the roughness arises from a two-dimensional random process. As with polycrystalline materials, a full description of rough surfaces would include a description of the height correlations across the surface. Frequently a Gaussian correlation is assumed, and this has the important property (Rice 1944, 1945) that it can be expanded near the origin in even powers of distance. In what follows, we shall assume that the correlations are negligible, and that the surface roughness may be described by a root mean square height h.

The scattering of ultrasound from rough surfaces is still the subject of much investigation. Current approaches often draw analogies with the scattering of electromagnetic waves from rough surfaces, as treated by, for example, Beckmann and Spizzichino (1963) and Bass and Fuks (1979), and full treatments of the elastodynamic problem are available only for the case of regular sinusoidal corrugations of the surface. The latter case has been studied by Fokkema and van den Berg (1977, 1980) and Fokkema (1980, 1981) by solving numerically a singular integral equation, and approximately by Roberts, Brind and Achenbach (1981) using the Kirchhoff approach. It is clear from their results that the mode conversion which occurs with elastic waves introduces effects which cannot be explained by a scalar wave theory, although for directions close to the specular reflection direction a scalar model gives a reasonable approximation provided that the surface is sufficiently smooth. For randomly rough surfaces Maradudin and Mills (1976) derived a theory based on Green functions, which could be solved in the limit of small surface roughness h and small slopes of the rough surface. The small slope approximation is also common in acoustic wave scattering calculations, in which the rough surface is treated as if it were made of a large number of plane facets (Biot and Tolstoy 1957; Novarini and Medwin 1978; Medwin and Novarini 1981). The other assumptions that are inherent in most treatments of rough surfaces is that there is no shadowing of the surface by itself, and that multiple scattering may be neglected. Liszka and McCoy (1982) have developed an integral equation approach, the first term of which gives the Kirchhoff result: higher terms give shadowing and multiple reflection effects. So far, their approach has only been applied to acoustic scattering. Variational methods have been used by Meecham (1956) and Heaps (1957) for periodic surfaces and by Lynch (1970) and Hart and Farrell (1977) for randomly rough ones. For a recent survey and extensions to the theory, see Ogilvy (1987).

For the scalar wave problem, Beckmann and Spizzichino (1963) have shown that, within the Kirchhoff approximation, the scattering may be divided into coherent and incoherent parts. Both parts contain an exponential dependence on the root mean square amplitude of the roughness, h, but the incoherent part contains in addition a dependence on the correlation length of the roughness. The intensity of the coherent term is given in terms of the Rayleigh parameter g

$$g = k^2h^2(\cos\theta_i+\cos\theta_r)^2 \tag{4.8}$$

for a wave of wavevector k incident at an angle θ_i and scattered at an angle θ_r. When mode conversion takes place, from an incident wavevector k^i to a reflected wavevector k^r, equation (4.8) may be generalized to

$$g = h^2(k^i\cos\theta_i+k^r\cos\theta_r)^2. \tag{4.9}$$

If the amplitude scattered by a perfectly plane surface is R_0, then the intensity scattered from the rough surface will be

$$I = |R_0|^2e^{-g}. \tag{4.10}$$

It has been shown by de Billy *et al.* (1980) that the coherent scattering dominates at low frequencies, up to about $kh = 0.5$, which corresponds to roughness heights on the scale of 0.1 mm for typical ultrasonic inspection systems working with frequencies in the MHz range.

4.6.3 Partially transparent cracks

As in the case of scattering from rough surfaces, the existing theories either treat unrealistically regular structures exactly or use approximate theories to model irregular structures. For tight fatigue cracks we expect there to be some interaction between the faces especially if the crack is under some compressive stress.

The most significant departure from all the phenomena treated so far is that this effect is likely to be nonlinear, because the contacts will be stiffer in compression than in tension (Richardson 1979). The nonlinearity has also been noted by Achenbach and Norris (1982). Haines (1980) developed a model of the transmission and reflection of compression waves at normal incidence on a tight fatigue crack. The model was based on earlier work of Kendall and Tabor (1971). In Haines' model the transmission of ultrasound by cracks includes the effects of surface morphology, the load under which the crack was grown, the applied stress, the flow stress of the material and the frequency of the incident wave. However, the results obtained are only for normal incidence which means that mode conversion effects are included. Temple (1985) has extended the method to cases of oblique incidence. The basis of the method is that contact is through small asperities, and that the asperities generate additional displacements both normal and tangential to the boundary. These additional displacements occur because of the stress transmitted by the passing elastic wave and depend on the nature of the contact, but are taken to be fully responsive to the driving force. Thus there is no relaxation time for the displacement induced by the passing stress wave and the surface responds immediately to any stimulus. Based on the concept of a surface stiffness, which differs from the elastic properties of a bulk material, Kendall and Tabor (1971) showed that the local deformation around the region of contact between an asperity on one surface and an otherwise flat surface is given by w where:

$$w = \frac{W}{\beta Y \bar{r}}. \tag{4.11}$$

Here W is the load on the contact, Y is Young's modulus, $\bar{r}$ is the radius of the contact and β is a constant lying in the range $\pi/2(1-\nu^2)$ to $2/(1-\nu^2)$ for a material with Poisson's ratio ν. The precise value of β depends on the shape of the asperity. Steel has a Poisson's ratio of about 0.28 giving β a range of values between about 1.7 and 2.2.

If a static load is applied to the crack then the faces will be in contact over discrete areas. The contact of these asperities can be described by elastic theory or, if the applied load is distributed over relatively small amounts of the total surface, by plastic theory. It is reasonable to assume that the load per contact is sufficient to create plastic deformation locally at each asperity when the crack faces first make contact. However, we shall allow elastic deformation only on subsequent transient loading of the crack unless the applied load becomes greater than the load at which the crack was grown. The static load does not appear directly in the equations of motion of the surface because it is a constant. However, as a stress wave passes through the interface the stress is transmitted by the asperities. As this happens, the asperity moves in and out of the surface into which it is indented by the applied static stress, and the amount of displacement which is created by the additional stress of the transmitted wave is modified because of the geometry of the asperity. Thus instead of solving the boundary equations for continuity of $u_x, \tau_{xx}, u_z, \tau_{xz}$, as in the case of a perfectly flat boundary, we solve a system of equations for $u_x^\dagger, \tau_{xx}, u_z^\dagger, \tau_{xz}$ where the displacements in the material on the side of the interface into which the wave is transmitted are modified by a perturbation which depends on the roughness properties of the interface. For the acoustic case Haines (1980) showed that the modification required is given by:

$$w = \frac{\tau_t \pi \bar{r} \sigma_m}{S\beta Y} \tag{4.12}$$

where τ_t is the stress transmitted at the interface, $\bar{r}$ is the mean radius of contact of the asperities, σ_m is the flow stress of the material, Y is its Young's modulus, and S is the applied load on the interface. The analysis of the elastic case proceeds by assuming that additional displacements are created by the passage of a stress wave. The approximation of modifying the displacements but not the stresses would be exact if the extra displacements around the contacting asperities were constant. The two extra displacements are u_x^t normal to the boundary and u_z^t parallel to the boundary and are given approximately by:

$$u_x^t = \frac{\tau_{xx}^t \pi \bar{r} \sigma_m}{S\beta Y} \tag{4.13}$$

and

$$u_z^t = \frac{\tau_{xz}^t \pi \bar{r} \sigma_m}{S\beta \mu} \tag{4.14}$$

where μ is the shear modulus. These terms are the additional perturbations inserted into the boundary conditions to give $u_x^\dagger$ and $u_z^\dagger$ on the side of the boundary into which the transmitted stress wave passes. Here $u_x^\dagger$ and $u_z^\dagger$ are given by

$$u_x^\dagger = u_x + u_x^t \tag{4.15}$$

$$u_z^\dagger = u_z + u_z^t. \tag{4.16}$$

Their effect is to build into the solution a term dependent on the frequency of the stress wave, and the magnitude of this term depends on the physical properties of the interface. It is not possible to write down an expression in closed form for the reflection and transmission coefficients for arbitrary angles of incidence, but for normal incidence of compression waves the result for the transmission coefficient T^{pp} is

$$T^{pp} = \frac{2}{\left[2+\frac{i\omega\pi\sigma_m\tilde{r}\rho c_p}{S\beta Y}\right]}, \tag{4.17}$$

which can also be written as

$$T^{pp} = \frac{2}{\left[2+\frac{ik_p Yw}{\tau_t}\right]}, \tag{4.18}$$

where k_p is the wavevector of the incident compression wave, and w is the perturbation introduced by contact of asperities. This expression shows how the various terms affect the transmission across the interface: increasing transmission with increasing applied stress S; and decreasing transmission with increasing frequency ω. The general results of this theory are confirmed by experimental work by Whapham, Perring and Rusbridge (1985).

Baik and Thompson (1984) have recently reviewed Haines' and other models of the contacting crack. They compare the expression for the transmission by a periodic array of cracks, as derived by Thompson and Fiedler (1983) using a quasistatic approximation, with an exact theory due to Angel and Achenbach (1985a,b). The functional form of the transmission coefficient at low frequency agrees with equation (4.17), and the agreement with the exact result is good at low frequency. It is possible to incorporate the effects of partial transparency into the finite difference model (Punjani and Bond 1986) and the basis function integral equation method (Visscher 1985), but detailed comparisons with other methods are not yet possible.

CHAPTER 5
Testing of pipelines

Nondestructive testing may be applied to pipelines, as it may to any system, for different reasons at different stages of their existence. At the factory, if pipes are to be made by rolling flat plate into cylinders, the plate is scanned for defects. The seam weld should then be checked before the pipe leaves the factory. When the pipe is installed in the system, girth welds are made between pipe sections, and quality control of these joints is necessary. Once the system is in service, it may be damaged by external impact, corrosion, fatigue, or by combinations of these. The effects of defects at each stage can be profound as regards the condition of the pipeline; in extreme cases they can lead to slow leakage or even catastrophic breakage of the pipeline system.

Inspection of a piping system when it is in service may be hampered by many circumstances. If the pipe forms part of a national transmission system it is likely to be buried below ground, and will therefore be inaccessible except from the inside. If it forms a key link in a national grid, it may not be possible to take the line out of service for inspection. This means that any inspection device must be capable of operating safely in a stream of gas under high pressure. Furthermore, the points at which an inspection device may be inserted in the line are likely to be few and far between: this means that it will not be possible to control the inspection through an umbilical link. As high pressure pipes are almost certain to be made of steel, radio control and data transmission is not possible.

Figure 70

A schematic diagram of the British Gas elastic wave pipeline inspection system, showing the two vehicles moving from left to right under the action of the gas stream on the cup rubbers.

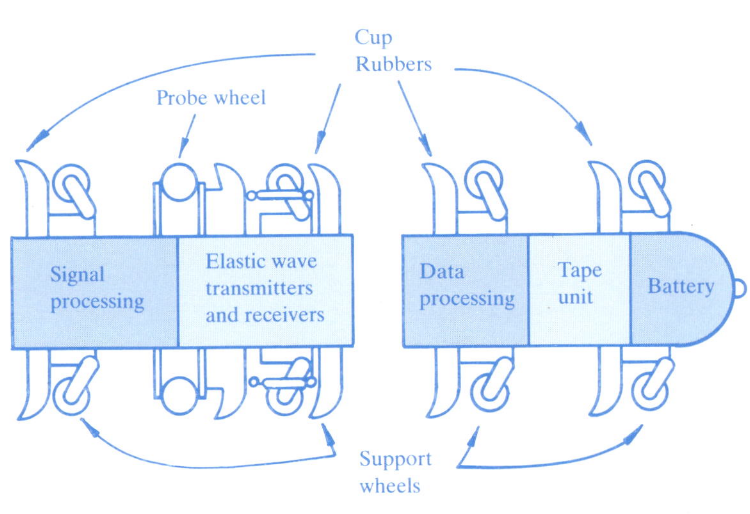

Figure 71

The British Gas plc elastic wave testing system for the on-line inspection of natural gas pipelines. The cup rubbers are the red rings encircling the vehicle, and the support wheels and the wheel probes for injecting the elastic waves are clearly visible.

These, then, were the conditions to be met by Harwell when considering how ultrasonics could be applied to an on line inspection system. A vehicle was required which could move along a pipe 900 mm in diameter in a stream of high pressure natural gas, injecting elastic waves into the pipe wall, detecting the resulting signals, and storing them on tape for subsequent analysis. The vehicle must be capable of operating at a speed close to that of the gas flow, and be capable of inspecting many kilometres of pipe at a single pass. In fact a pair of vehicles is used, one of which provides the power and carries the tape recorder while the other handles the ultrasonic part of the inspection.

A schematic view of the system is shown in Figure 70, and Figure 71 shows the vehicle being prepared for an inspection run. Here we shall consider only the ultrasonic aspects of the system. The technical problems involved in mechanical design, tape recorder development, on-board data reduction and off-board data processing would fill several more books, but some are worthy of mention. The vehicles had to be strong enough to survive running at speeds up to 36 km/hr, greatly in excess of normal inspection speeds. They had to be able to pass round sharp bends in the pipework, over the girth welds between pipes, and through sections where the internal diameter of the pipes may suddenly decrease by up to 50 mm. This last requirement allows for the specially strengthened sections of pipe which occur at right angled bends.

First we shall describe the aims of the inspection system, and this will be followed by a short description of the geometry chosen. Finally, we shall show how a composite model of the elastic wave behaviour in the pipe wall was used to model the responses to be expected from the features to be found in the pipes.

5.1 Detection requirements

A pipe carrying gas at high pressure sustains a large hoop stress in the pipe wall. If the wall is locally weakened by the presence of a defect, failure is possible.

The failure depends on both the strength of the pipe, as measured by its yield stress, and its toughness. Toughness is a measure of the amount of energy absorbed in fracturing a material, and depends on the mode of fracture – brittle or ductile. In brittle fracture the crack propagates at the speed of elastic waves in the solid: as this is much greater than the speed of sound in the gas the only energy available for crack propagation is the elastic energy stored in the pipe wall under hoop stress. Brittle fractures, once initiated, will propagate right through the structure unless arrested by some gross change in material properties, as long as enough elastic energy is available. If the material of the pipe wall flows plastically, however, the stress at the crack tip can be relieved and brittle fracture avoided. There is a characteristic temperature T_c above which the fracture mode is ductile and below which it is brittle, and it is important that pipelines should be operated above their T_c values. In cases where this requirement was not satisfied, brittle fractures extending over ranges of several miles occurred (O.G.J. 1960). The drop weight tear test (DWTT) was devised as a means of determining T_c for pipeline steels (McClure, Duffy and Eiber 1965). It later became clear that very long ductile fractures could also occur (Maxey 1974). Such fractures propagate more slowly, at a few hundred metres per second, which is slower than the speed of sound in the gas inside the pipelines. This means that energy can be supplied by the gas to drive the crack forwards: the mechanism seems to involve the pressure of gas on the flaps behind the opening crack helping to tear open the crack tip. The process was studied in detail at Battelle Laboratories (Maxey 1974) and by British Gas (Poynton 1974), using models which involve the decompression characteristics of the gas. The detailed approaches are slightly different, but both models have been quite successful at predicting the propagation and arrest of ductile fractures. They are extensions to the science of fracture mechanics as described, for example, by Knott (1973).

Armed with such models, one is able to predict what the effect on the pipeline of defects of various sizes will be. The first important concept is the leak–break boundary. If a crack is short in the axial direction, but penetrates a large fraction of the pipe wall, it is likely to propagate through the wall to form a leak which will not extend further axially. Longer defects of significant depth are likely to fail by rupture, and are therefore more critical. The leak–break boundary depends mainly on the tensile properties of the material, so once the operating characteristics of a pipeline are specified in terms of the ratio of the operating pressure to the yield stress of the pipe material, the theoretical leak–break boundary is known. For much of the high pressure gas main in service in the United Kingdom the boundary lies at about 100 mm.

Figure 72

Detection requirements of the Pipeline Inspection System. The crack dimensions for which pipeline failure would occur are shown.

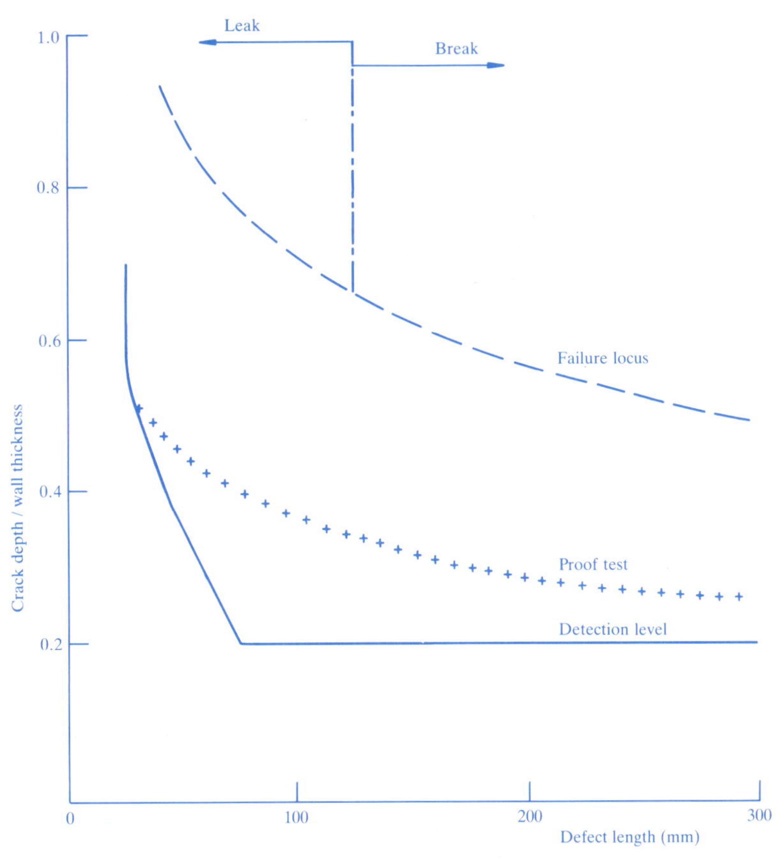

The next point to be determined is the depth of penetration at which the pipe wall becomes unstable. Unless the toughness is very low, the failure is again plastic and depends on the tensile strength of the pipe steel. This defines the critical depth of the defect. In practice, corrections must be made to allow for very short defects and defects of varying depth penetration. Figure 72 shows the combinations of defect lengths and wall penetration that would result in failure with the leak–break boundary marked. Figure 72 shows that cracks down to about 20 per cent of the pipe wall thickness are to be detected, but that as their length decreases the significant depth increases to 40 per cent or more of the wall thickness.

Figure 73

The interactions between defects when they are situated close to one another in the longitudinal direction.

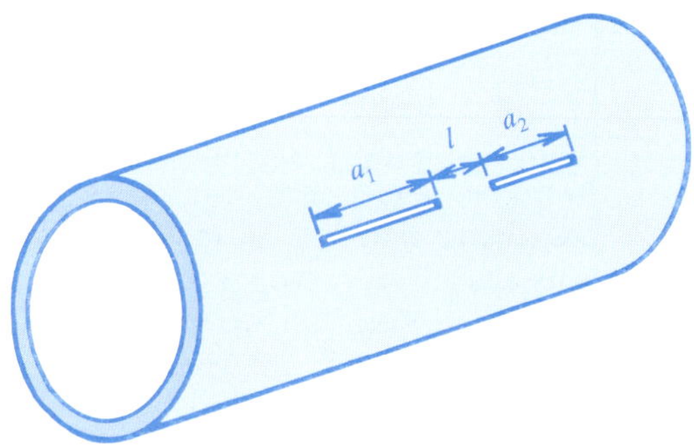

When several defects are close together, their stress fields interact and for the purposes of determining the pipe failure characteristics they may have to be combined. These interactions have been studied in detail by British Gas at the Engineering Research Station. Figures 73 and 74 show typical situations. The rules specified for combining crack-like defects aligned parallel to the pipe axis may be stated quite simply. If the cracks lie in the same radial plane, their lengths are a_1 and a_2, and the separation between their ends is l, then when l is less than both a_1 and a_2 they are treated as though they were one crack of length a_1+a_2+l. If, on the other hand, a family of cracks has circumferential spacings c_i less than the pipe wall thickness h (Figure 74), then it is to be treated as a single crack with overall length equal to the total length of the family, d.

Figure 74

The interactions between defects when they are situated close to one another in the circumferential direction.

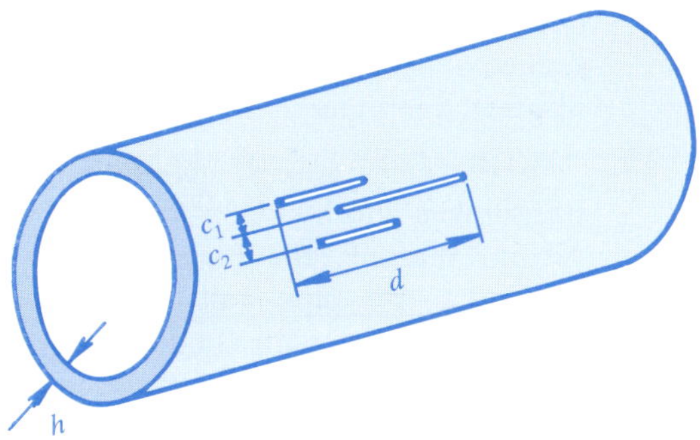

5.2 The design of the testing system

The metallurgical properties of the high-strength steel used in high pressure gas pipelines result in a medium which is essentially elastically isotropic, and which attenuates elastic waves very slightly in the megahertz range. This means that elastic waves can be propagated satisfactorily over ranges of the order of metres. The pipes in service in the United Kingdom are wrapped with pitch or epoxy, to protect them against corrosion: this has the advantage that the medium which surrounds the pipe is quite well characterized, but leads to significant loss of amplitude of the waves at each reflection from the outside of the pipe wall, as described below.

Figure 75

For a pair of wheel probes on a segment of pipe wall, the figure shows two transducers, (a) and (b), inspecting a reflector in pulse echo mode.

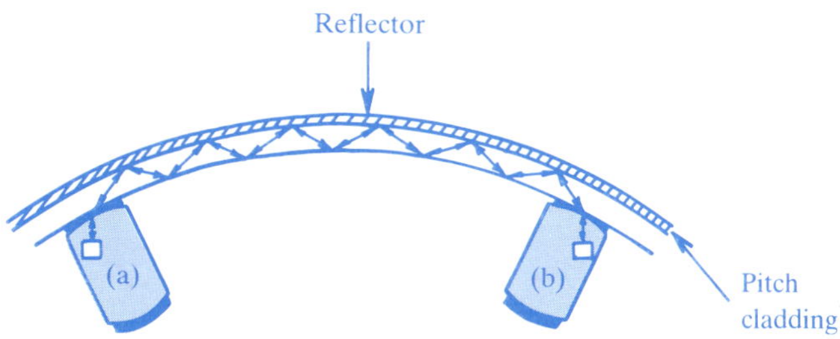

The major stress in the pipe wall is a hoop stress arising from the gas pressure, so the cracks which the system is required to detect will grow preferentially in a radial plane, perpendicular to the pipe wall and parallel to the pipe axis. The most promising configuration, therefore, is a system in which elastic waves are injected so as to travel circumferentially around the pipe, and are detected as they are reflected from the cracks. This is the pulse echo technique, and is illustrated in Figure 75.

In the early stages, generation of elastic waves by piezoelectric transducers and by electromagnetic means were both considered. For various reasons, the former method was selected. The signals are coupled to the pipe wall by mounting the transducers inside wheels which are filled with fluid. The wheels have rigid rims, which are coupled to the pipe steel through flexible tyres. The passage of the elastic waves to the steel is therefore through the multilayered system of transducer, fluid, rim, tyre, steel, with significant losses at each interface. The worst losses occur between the tyre and the steel, and the coupling at this interface may be improved by using a fluid layer or by applying a high load to the wheel to press the tyre firmly against the pipe wall.

As shown in Figure 75, each reflector is inspected by beams injected clockwise and anticlockwise around the pipe wall. In addition, each transducer in a pair is able to detect the direct signal from the other as well as its own reflected signals. This allows compensation for the effects of variations in the efficiency of injection of elastic waves into the pipe.

5.3 Choice of wave mode and angle

When selecting a nondestructive testing technique for pipeline inspection, it is necessary to take into account the detection requirements, the losses in transmission, and the ease with which the input signal can be generated. These constraints allow one to choose the optimal wave mode from the range of possibilities: bulk compression, bulk shear (with a choice of polarization), surface waves or Lamb waves.

Surface waves can be discarded immediately. They are confined to the inner surface of the pipe wall, where the transducers are, yet the hoop stress on the pipe is greatest at the outer surface, so fatigue cracks would grow inwards from there. Stress corrosion cracking would normally be initiated at the outer surface of a gas pipeline. Surface waves, then, would only detect the most dangerous defects when they had already penetrated the pipe wall.

Lamb waves appear promising, and the calculations described in Chapter 4 show that they are sensitive to surface-breaking defects. The low order modes are only suitable for sizing shallow cracks (as Figure

68 shows) because the reflected amplitude saturates too rapidly with crack depth. Their strong dispersion raises further difficulties. To select a single mode near an extremum of the dispersion curve, to reduce the effects of dispersion, a long pulse is required. This results in poor resolution around the pipe circumference, and the defect cannot be located to an accuracy greater than several pipe wall thicknesses. On the other hand, if a short pulse is used it will be spread by dispersion and resolution will again be low.

Only the bulk modes remain to be considered. The most attractive mode is the shear wave polarized parallel to the pipe surface (SH wave). Such waves would propagate through the pipe wall without losses arising from mode conversion, and would also be diffracted from a radial crack without mode conversion. Unfortunately such waves cannot be generated in the pipe wall by a normal stress on the surface, which is all that is available from a wheel-mounted transducer. Compression waves, incident at an angle on the inner (steel–gas) and outer (steel–cladding) surfaces of the pipe wall, will suffer loss through mode conversion to shear waves at each reflection. It is not possible to cover a practical circumferential range with such waves. Shear waves are also subject to mode conversion, but above the critical angle they are totally internally reflected at a stress free surface. There is always some loss of intensity into cladding on the outside of the pipe wall at each reflection.

This suggests that the best coverage of the pipe wall would be attained by using shear waves at very high angles, to minimize the number of reflections for a given range. Against this must be counted the fact that high angle shear waves are always accompanied by surface waves. This was shown in the calculations in the previous chapter, and also in experimental work where the ratio of surface wave to shear wave was found to rise rapidly at high angles. Surface waves will be strongly reflected by surface blemishes which are too weak to have any significance for the strength of the pipes. This effect is so strong that one experimental system using high-angle shear waves returned strong signals from paint runs on the inside surface of the pipe. This enormous quantity of spurious signals would place an intolerable burden on the data collection and data processing systems. On the other hand, too low an angle is undesirable for another reason. If a wave propagates through the pipe wall at an angle of $(\frac{1}{2}\pi-\theta_c)$ to a radius, where θ_c is the critical angle, it will be incident at θ_c on a radial crack: this is near the minimum reflectivity from a plane surface. The wave angle chosen must be a compromise, then, between strong reflections from defects over a wide range and unwanted reflections from surface features.

5.4 Modelling of pipeline defects and detection

When a nondestructive testing system for a complicated problem is being developed, experimental work frequently produces unexpected results. In the course of Harwell's development of an on-line inspection system for gas pipelines, several such cases came to light.

It is vital to study these results in detail, to ensure that they are not common situations which could give rise to enormous numbers of spurious indications on the one hand, or obscure dangerous defects on the other. Examples of the first kind were strong reflected signals from features inside the pipe wall. Some were examined destructively, and found to consist of laminar inclusions of clouds of alumina particles, manganese sulphide, calcium silicate, and one example was found of a large copper inclusion. Many of these inclusions originate during manufacture either as a result of accidental uptake of flakes of refractory dressing from moulds, or from the addition of alloying materials at the wrong temperature or in the wrong form. Other unexpected results were obtained when the pitch layers with which the pipe was wrapped for protection when buried had been applied in such a way as to give a layered pitch cladding. Finding and explaining the unexpected cases is only part of the problem: it is then necessary to ensure that the inspection system can cope with similar cases. To do this adequately it is necessary to have a detailed model of the inspection system, which can explore a larger range of cases than are available experimentally.

The most complete modelling of the interactions of ultrasound is undoubtedly that afforded by finite difference models. The methods applied in such calculations have been described above (Chapter 3), and applications to elastic wave generation and scattering have been given elsewhere in Chapter 4. The very completeness of the finite difference model is, however, a drawback. The numerical solution of the elastic wave equations at every point in a grid does not allow one to apply knowledge of the dominant processes: the computer has to calculate even the propagation of a plane wave through a homogeneous space as though such propagation had never been seen before. This, of course, means that the calculation of processes in the far field of a transducer becomes very expensive. As an example, the generation of a shear wave at the surface of a plate 10 wavelengths thick and its propagation to and scattering from a defect at a range of 40 wavelengths requires a finite difference mesh containing about 100 000 points, and takes a whole weekend to run on a Digital Equipment Corporation VAX 11/780 computer. Although this time would be reduced to tens of minutes on a supercomputer such as the Cray series, additional problems with the finite difference method occur when the three-dimensional nature of the problem is considered, and when model reflectors such as rough cracks, or clouds of small alumina particles are to be studied. Such models have to be treated by a model rather than an explicit calculation because, for example, it may be necessary to include averaging processes for rough surfaces.

To allow complete analysis of the ultrasonic scattering process in a section of pipe wall between two probes, some simplifications must be made. The aim of a ray tracing model is to synthesize a model from the elements of ultrasonic behaviour which are well understood. The generation, propagation and scattering of elastic waves are treated separately, using the well established theories which have been reviewed in this book, and it is then relatively easy to handle the geometrical problems involved in joining them smoothly.

5.5 The ray tracing model

The calculation of the signal seen in the pipeline inspection system may be broken down into three components: the generation (and detection) involving the modelling of the transducers; the propagation of the ultrasound through the pipe wall; and the scattering by the reflectors.

5.5.1 Generation – the transducers

There are several effects to be taken into account in modelling the behaviour of the transducers. As the transducers are the inputs and outputs of the model, all the effects of the driving pulse, electrical filtering and resonant behaviour of the transducer elements are treated under this heading. The geometrical shape of the beam from the transducer must be modelled, and so must the shape of the input pulse.

As the transducer is, in the current design of pipeline inspection vehicle, enclosed in a wheel filled with fluid, the beam profile may be modelled by using expressions for a source of ultrasound in a fluid. This is a classical problem and the pressure in the far field at an angle θ from the axis and a distance z from the centre of a transducer of radius a excited by a continuous wave of frequency ω is

$$p(z,\theta,t) = \frac{i\mathrm{P}_0 ka^2 e^{i(\omega t - kz)}}{z} \frac{J_1(ka \sin\theta)}{ka \sin\theta}, \tag{5.1}$$

where k is, as usual, the wave number ω/c. The far field is generally taken to lie in the range

$$z > \frac{4a^2 - \lambda^2}{4\lambda}, \tag{5.2}$$

for a wavelength λ. In the near field there are complicated variations of amplitude and phase (Beaver 1974, Foster and Hunt 1978), but pulsed operation gives a much smoother distribution of power than does continuous wave operation.

The Bessel function J_1 gives rise to side lobes in the beam pattern, but most of the energy is contained within the central lobe. The profile of equation (5.1) may be replaced by a Gaussian with reasonable accuracy. Akoev (1975) has noted that the fraction of the total energy predicted by equation (5.1) to fall within the central lobe, which is 82 per cent, corresponds to the 82 per cent of the energy found within a Gaussian beam within two standard deviations of the centre. The Gaussian beam approximation leads to the following expression for the beam profile:

$$p(z,\theta,t) = \frac{ip_0}{z} e^{i(\omega t - kz)} e^{-\theta^2/2\sigma^2} \tag{5.3}$$

for small angles θ. If the Gaussian and Bessel function profiles are related by equating their widths at half height, one obtains

$$\sigma = \frac{2.2}{2ka\ln 2}. \tag{5.4}$$

In practice, the radius a of the transducer is not equal to the geometrical radius, as it is affected by details of the fixture of the piezoelectric crystal in its mounting. The effective radius may quite readily be determined by experiments in a water bath.

Although the beam profiles given in equations (5.1) and (5.3) are symmetrical, it is important to realize that the beam in the pipe wall will not be symmetrical. This is because the transmission of elastic waves from a fluid into steel varies with angle, as shown in Chapter 2. As a result the beam in the steel is biased to high angles. This transmission factor has recently been incorporated into models of ultrasonic inspection by Achenbach, Lin and Keer (1986).

In the time domain, the transducer often is driven by a voltage edge, which is smoothed out by the electronics and by the fundamental resonance of the transducer. The pulse shape is assumed to have a sharp rise followed by an exponentially damped sinusoidal oscillation. The analytic form is taken to be, for a pulse beginning at time t_0,

$$\phi(t-t_0) = (t-t_0)^n \cos\omega_0(t-t_0) \quad e^{-\alpha(t-t_0)} \tag{5.5}$$

and the parameters n, ω_0 and α have to be determined by comparing the pulse shape with experimental pulse shapes.

To increase the flexibility of the input pulse shape, it may be modified in two further ways. A multipulse form may be used, in which two pulses, each of the form of equation (5.5) but separated by a small time, make up the overall pulse. Alternatively, the effect of several different decay times and resonant frequencies may be modelled by making ω_0 vary during the pulse. By allowing the frequency to decrease, so that we may write

$$\omega_0(t) = \omega_0 - \omega_1 t, \tag{5.6}$$

we have found it possible to create pulses which closely resemble experimental pulses. As the experimental pulses are themselves variable, depending on small details of the transducer construction, the optimal parameters for the model pulse were not fitted by numerical procedures. It was found that the fitting was best carried out by referring to time and frequency domain graphs of experimental and model signals.

The varying frequency content of a beam as a function of angle with respect to the beam centre is well known to practitioners of nondestructive testing, but it is often ignored in theoretical models which use a single frequency. The change of frequency can be quite marked. In the Time of Flight method (Silk 1977) a signal may pass from one surface-contacting transducer to another either by reflection from a defect or as a lateral wave creeping along the surface. The lateral wave, which is transmitted and reflected in the edges of the beams, can have a mean frequency as low as half the mean frequency of the main beam. The origin of this effect is clear from equation (5.1). The first zero of the Bessel function beam profile is given by

$$\theta = \sin^{-1}\left(\frac{3.8}{ka}\right), \tag{5.7}$$

so that the beam spread is approximately proportional to the inverse of the frequency. This means that the edges of the beam are biased to lower frequencies. In essence, this is because the transducer looks, for low frequencies, more like a point source with its uniform angular spread than like a large transducer with highly directional beam. This effect may be included in a fairly simple way, by adjusting the centre frequency of the pulse. Thus ω_0 in equation (5.5) is made a function of the beam angle θ. A suitable approximation may be found by evaluating the expectation value of the frequency at an angle θ to that at the centre of the beam for a Gaussian pulse

$$g(t) = e^{-\frac{1}{2}\Delta^2 t^2} e^{i\omega_0 t}, \tag{5.8}$$

which gives

$$\frac{<\omega>_\theta}{<\omega>_0} \approx \left(1 + \frac{\sqrt{2}\Delta + 2\pi\omega_0}{\sqrt{2}\Delta + \pi\omega_0} \frac{\Delta^2}{2\Sigma^2}\theta^2\right)^{-1}. \tag{5.9}$$

Again, one may determine the constants from experiment, and if the centre of the beam is at θ_0 write

$$\omega_0(\theta) = \frac{\omega_0(\theta_0)}{1 + P_\omega(\theta - \theta_0)^2}, \tag{5.10}$$

where P_ω is determined from the experiment.

It should be noted that when a pulse is examined using the experimental arrangement of reflecting a signal from a plane surface using the same transducer as transmitter and reflector the signal seen is the convolution of the pulse with itself (assuming that the response function of the transducer as a transmitter is the same as that as a receiver).

5.5.2 Propagation – the geometrical problem

The geometrical problem is to find the possible paths for a ray from a point on the inside surface of the pipe wall to an arbitrary point within the pipe wall. The second point may be a defect, at any point within the wall, or a receiving transducer on the surface of the pipe wall. The results given below are for a two-dimensional case in which the two points lie in one plane perpendicular to the pipe axis. We refer to the path between the two points as a series of skips: the path is defined by its angle with a pipe radius.

Figure 76

The geometry of the ray tracing calculation, showing a ray from the transducer at T which makes one whole skip and rather more than one half skip on its way to a reflector at P.

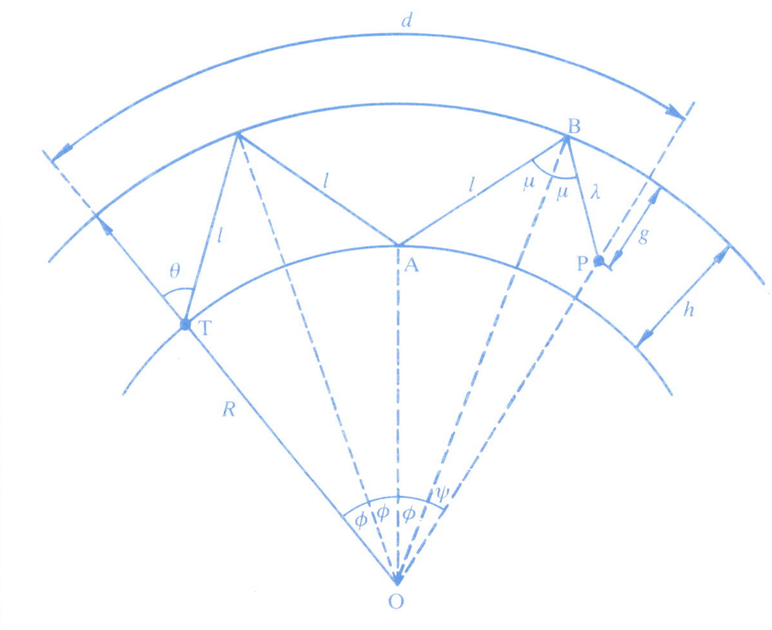

Figure 76 shows such a ray path, between a transducer T and a point P on a defect. That point is at a depth g from the outer surface, and the thickness of the pipe wall is h. There are n whole skips each of length l and one partial skip of length λ. We may reach the point P by a whole number of skips plus or minus a partial skip. This means that we allow only an odd whole number of skips (n = 1,3,5 etc) and the partial skip is always measured from the outer pipe surface. The outer radius of the pipe is R. The angle between the ray path and the pipe radius is θ. Each of the full skips subtends an angle ϕ at the axis of the pipe, and the partial skip of length λ subtends an angle ψ.

In Figure 76, the pipe axis has been labelled O, the points at each end of a particular whole skip are labelled A on the inner pipe surface, B on the outer surface. Then, in triangles Δ OAB and Δ OBP the cosine rule may be used to give

$$(R-h)^2 = R^2+l^2-2Rl\cos\mu \qquad (5.11)$$

$$(R-g)^2 = R^2+\lambda^2-2R\lambda\cos\mu \qquad (5.12)$$

so that

$$\lambda(R^2+l^2-(R+h)^2) = l(R^2+\lambda^2-(R-g)^2). \qquad (5.13)$$

Thus for whole skips, when $g = 0$ or $g = h$, we obtain

$$l_n = \sqrt{2R(R-h)\{1-\cos(\frac{d}{nR})\}+h^2} \qquad (5.14)$$

and

$$\theta_n = \cos^{-1}\left\{\frac{h-R\left\{1-\cos(\frac{d}{nR})\right\}}{l_n}\right\}. \qquad (5.15)$$

Some similar results have been given by Johnson (1985). For partial skips, where the defect is not on either of the two pipe surfaces, a numerical solution of the problem is used. This involves the use of the Newton–Raphson method to solve a non-linear equation for the segment lengths l_n and λ_n.

The form of the signal is, then, a superposition of signals arriving at times determined by these path lengths. Each path corresponds to a different angle relative to the pipe wall, and therefore to a different part of the beam profile. As a result, the amplitude and the frequency content will vary from skip path to skip path. This makes it hard to use information on the frequency content of the scattered signal to determine the structure of the scatterers, as it is necessary to correct the spectrum for range-dependent effects.

5.5.3 Models of scattering

From the description of elastic wave scattering given in Chapter 3, it is clear which theories are appropriate for the modelling of the testing system. For small scatterers, we use the quasistatic approximation. For crack-like defects, we combine the ray theory and Kirchhoff approximations, using the former when only a diffracted signal is expected and the latter close to the specular direction. This combination of models gives an accurate picture of the scattering from a large range of defects. In applying each theory, the effects of the particular scattering process on the frequency content of the signal are included in a simple way, by inserting a filter with the frequency behaviour of the leading term in the theory.

5.5.4 Constructing the signal

The components described above have to be assembled correctly to predict the resultant signal. The effects of phase changes on reflection have been considered in Chapter 2, and they must be included here. If there were no phase changes, the composite signal could be built up very simply in the time domain. Phase changes, however, introduce rather subtle changes of shape into the waveform. For example, a delta function input pulse

$$\psi(\mathbf{x},t) = \delta(\mathbf{k}.\mathbf{x}-t) \tag{5.16}$$

undergoing a phase change of α will emerge as a pulse of the form

$$\psi_\alpha(\mathbf{x},t) = \cos\alpha\ \delta(\mathbf{k}.\mathbf{x}-t) - \frac{1}{\pi}\sin\alpha\ P\,\frac{1}{\mathbf{k}.\mathbf{x}-t}. \tag{5.17}$$

Thus the pulse acquires both a tail and a precursor, as shown in Chapter 2.

For computational purposes, the delta functions are replaced by functions which are numerically less violent. The terms of equation (5.17) are replaced as follows:

$$\delta(\mathbf{k}.\mathbf{x}-t) \rightarrow \frac{1}{(\mathbf{k}.\mathbf{x}-t)^2-\varepsilon^2} \tag{5.18}$$

and

$$\frac{1}{\mathbf{k}.\mathbf{x}-t} \rightarrow \frac{\mathbf{k}.\mathbf{x}-t}{(\mathbf{k}.\mathbf{x}-t)^2-\varepsilon^2} \tag{5.19}$$

where ε is typically chosen to be equal to one of the time units in which the signal is discretized.

Of course, the inclusion of phase changes requires Fourier transform techniques to be used to form the final signal, convoluting a response of the form of equation (5.17) with the input pulse shape. This leads to substantial increases in computing time, but is necessary if reliable results are to be obtained.

5.5.5 Theoretical signals

The best confirmation of the general principles of the ray model is obtained by comparing the predicted pulse shapes transmitted from one transducer to another with the experimental pulses. The signals should be compared in the frequency domain as well as the time domain, as the two presentations emphasise rather different features of the pulses. The parameters introduced in the previous sections, n, ω_0, α, P_ω and σ were varied to achieve good agreement between the model and experiment, but in no case was the value of the parameter far from the initial theoretical estimate. Numerical comparisons were made by passing the theoretical signals, digitized in the same way as the experimental ones, through the experimental data processing system: the agreement was excellent.

5.6 Applications of the modelling system

The reason for developing the modelling system was to be able to study, rapidly and reliably, a large range of testing conditions and defect types. A modelling system was used for several reasons. In a real pipeline one cannot readily introduce scatterers of arbitrary type: the development of the testing system rested heavily on theoretical models of the scatterers likely to be present.

The models were developed on the basis of metallurgical examination of some of the unexpected reflectors found in line pipe, and on artificially grown examples of stress corrosion cracks. The model could easily be adapted to study pipes of different diameters, where to rebuild the real testing vehicle would have been very expensive. Similarly, the effects of different pipe wall thicknesses could be analysed. The condition of the pitch cladding on the outside of the pipe could be varied in a controlled fashion, to examine the effect of the cladding on the ultrasonic signal. As the model signals were processed in exactly the same way as the real signals, the processing scheme itself could be checked, and tested for such unwanted effects as a dependence on the range from the transducers to the reflector or the condition of the cladding.

Throughout the development of the elastic wave testing system, the theoretical model was developed in parallel with the experimental vehicle. This ensured that at all stages a full understanding of the processes affecting the ultrasonic signals was available. In difficult cases, where the several conflicting effects were operating, numerical results were available to demonstrate that no new, unexpected, phenomena were being encountered.

CHAPTER 6
Summary and prospects

In recent years, the theory of elastic wave scattering has evolved rapidly as the demands of nondestructive testing have become more refined. This reflects a general increase in the whole area of reliability, which is a natural consequence of increased safety consciousness on the one hand, which makes a failure of certain large structures socially and politically unacceptable, and economic considerations on the other, which demand that plant shall not be closed down without convincing evidence that it poses a risk. This philosophy is embodied in the retirement for cause approach which has been developed in the work programmes sponsored by the Defence Advanced Research Projects Agency, United States Department of Energy and Air Force Wright Aeronautical Laboratories. The result is that far more detailed information about the scatterers seen in structures must be extracted from ultrasonic measurements.

The purpose of this monograph has been to show what theories are available for calculating the scattering by known reflectors. The aim has been to stress which theories are most useful for particular cases. Chapter 5 has, it is hoped, demonstrated that a theoretical model can be constructed which will compute pulse shapes from ultrasonic scattering which may be compared directly with experimental results, and used to interpret and extend them. One important application of reliable models is in assessing the ability to detect 'reasonable worst case flaws' (Coffey, Chapman and Hanstock 1982, Coffey and Chapman 1983). The position and orientation of the flaw that would be most difficult to detect are estimated, and the model is used to see whether the signal from such a flaw would exceed the detection threshold.

The theories described in this monograph represent the current state of the art in the direct problem, that is, the calculation of the output elastic wavefield when a known input field interacts with a known structure. The problem of inferring from measurements made from outside a body the detailed structure of its interior is commonly referred to as the inversion problem. It is clear that a very firm understanding of the direct problem is necessary (although possibly not sufficient) if the inversion problem is to be solved. In most cases the inverse problem may be simplified if it is known that only certain types of scatterer exist in a given structure. In the case of gas pipelines, it was at first assumed that only crack-like scatterers would be found: subsequent experiments disclosed metallurgical features of several kinds, which were analysed by destructive examination, and means had to be devised of distinguishing them from cracks. It seems likely that some such prior information will always be necessary in solving the inverse problem.

Throughout nondestructive testing it should be kept in mind that the process has a probabilistic element. In recent years extensive exercises have been undertaken to investigate this aspect. In the United

Kingdom, for example, specimens including defects have been examined by several groups using different techniques and the results have been analysed statistically to derive reliability estimates on the detection and sizing capabilities of the different methods (Crutzen, 1982). Much work has been devoted to coupling probabilistic fracture mechanics with models of detection reliability: the studies of Pressurized Water Reactor (PWR) pressure vessels by the Marshall study group (Marshall 1976, 1982) are particularly thorough. The aim is to devise a function $B(a)$ which represents the probability that a defect of size a remains in a structure. There are contributions to the population of such flaws from initial defects which went undetected during manufacture and from defect growth in service: the population is reduced by in-service inspection and repair. In Marshall (1976, 1982) it is suggested that $B(a)$ should have the form

$$\begin{aligned} B(a) &= \varepsilon + \gamma e^{-\mu a} \qquad & a < a_{crit} \\ &= 0 \qquad & a \geqslant a_{crit}. \end{aligned} \tag{6.1}$$

This shows that there is some low probability ε that a defect will be undetected, no matter how big it is, unless it is of such a size ($>a_{crit}$) that it causes leakage or rupture. This low residual probability represents, for example, human factors such as the omission of an inspection. The exponential term represents the increasing ease of detection of larger defects. In general, there will be some lower size threshold below which defects are not required to be detected. Cameron and Temple (1985, 1986) have used modelling techniques to evaluate the parameters in $B(a)$: μ is essentially a measure of the capability of inspection to distinguish critical from non-critical defects. An important conclusion of their work was that the usefulness of increasing the accuracy of ultrasonic sizing schemes is restricted unless parallel improvements are made in the likelihood of gross errors. Williams and Mudge (1985) stress the importance of design and fabrication: in welded structures the occurrence and severity of defects at these early stages limit the structural integrity, and increasingly accurate ultrasonic measurements are of limited value. A detailed survey of inspection reliability is to be found in the book by Silk, Stoneham and Temple (1987). The probabilistic aspects of the inversion problem have been treated by Richardson (1982) and Marsh *et al.* (1982) within the context of the Born approximation.

Much of the effort in underlying research in elastic wave scattering has, in recent years, been directed towards comparing very detailed experimental measurements of scattering from idealized (although not necessarily trivial) artificial defects with theoretical results on the same systems. The emphasis is now shifting towards more general problems, and the coming years may be expected to bring an increase in the modelling of complete systems, perhaps along the lines of the system described in Chapter 5. A committee of the American Society of

Mechanical Engineers recently surveyed the fields of elastic waves and quantitative nondestructive evaluation (Clifton 1985) and concluded that further progress depends critically on large-scale computations. Areas which are of particular interest include wave propagation in inhomogeneous and anisotropic media, scattering from rough and semi-transparent cracks, characterization of transducers, and the detection and identification of defects. New materials, such as ceramics, bring new problems. There is a need to extend calculations to three-dimensional problems. Certainly the problems tackled by theorists will continue to converge on the questions posed by safety analysts.

Glossary of terms

angle of incidence – the angle between the direction of a wave and the normal to a surface on which it impinges.
angle of reflection – the angle between the direction of a wave and the normal to a surface from which it is reflected. Except in the case of mode conversion *(q.v.)* the angle of reflection is equal to the angle of incidence.
angle of refraction – the angle between the direction of a wave and the normal to a surface from through which it passes.
anisotropy – the quality of having properties which vary with direction.
asymptotic method – method appropriate for treating a mathematical function when its argument is large: here applied to wavefields at long ranges.
attenuation – the loss in energy as a function of distance travelled by a wave.
backscattering – scattering *(q.v.)* through angles greater than 90°.
beam – a radiation pattern with a wavefront of limited width.
Born approximation – a form of perturbation *(q.v.)* expansion.
canonical problem – a problem which is typical of those arising in many diverse situations.
caustic – a surface which is the common envelope of many rays.
compression – a stress which tends to reduce the volume of a body without altering its shape.
compression wave – a wave in which the direction of particle motion is parallel to the direction of propagation of the wave.
coupling – the efficiency with which elastic energy is transmitted between two bodies.
creeping wave – a wave which follows a smooth surface rather than travelling along straight rays.
critical angle – incident angle of an elastic wave beyond which a specified refracted wave or mode-converted reflected wave no longer exists.
diffraction – the deflection of a wave as it passes the edge of a scatterer.
discretization division into small cells, as a prelude to finite difference or finite element *(q.v.)* calculation.
dispersion variation of wave speed with wavelength (used by other authors to denote scattering as a result of reflection from irregular surfaces).
eddy current – current induced in a conducting specimen by the application of a time varying magnetic field.
elastic constants – constants of materials which describe their response to applied stress.
energy flux – rate of flow of energy across unit area.

far field – region remote from a transducer or a scatterer, in which maxima and minima of intensity as functions of distance no longer occur.
finite difference method – method of solving differential equations by approximating derivatives by numerical differences between function values at a set of grid points.
finite element method – method of solving differential equations by approximating functions by interpolating functions within small regions, smoothly joined at the edges of the regions.
fracture mechanics – method of predicting the behaviour of bodies which contain stress-concentrating defects.
group velocity – velocity at which a pulse of energy propagates: may differ from phase velocity in a dispersive medium.
head wave – wave generated as a wave of higher speed moves tangentially to an interface.
homogeneous – not varying with position in space.
Huygens' principle – theory that each region of a wavefront acts instantaneously as a source of further waves.
inhomogeneous – varying with position in space.
insonification – is to sound or elastic waves as illumination is to light: bathing of a region in elastic wave energy.
interference – pattern of reinforcement and cancellation formed by the superposition of two or more wavefields.
inversion – inference of the details of a structure from measurements taken from outside it.
isotropy – the quality of having identical properties in all directions.
Lamb wave – a combination of compression and shear waves propagating in a plate, normally with a wavelength comparable with the plate thickness.
longitudinal wave – same as compression wave *(q.v.)*.
mode conversion – a process that changes the type of an elastic wave from compression to shear or *vice versa.*
near field – region close to a scatterer or transducer within which the wave pattern changes in a complicated manner, giving rise to maxima and minima as functions of range.
nondestructive testing – searching for internal or surface defects in materials without damaging the specimen being tested.
optical theorem – theorem relating forward scattering to total scattering cross-section, based on energy conservation.
perturbation method – means of calculating scattering when the scatterer differs only weakly from the host medium.
phase velocity – velocity of individual peaks and troughs of a wave.
pitch and catch – mode of testing in which two transducers are used, one as the transmitter and the other as the receiver.
polarization – direction of motion of particles in a wave, related to the direction of travel of the wave.
pulse echo – mode of testing in which one transducer is used as transmitter and receiver.

ray – path followed by a wave in the high-frequency limit: corresponds to the paths of geometrical optics.
Rayleigh wave – a wave confined to the free surface of a solid material, with an amplitude decreasing exponentially with depth.
reciprocity – relationship between the effect at a second point arising from a disturbance at a first point and the effect in the inverse situation.
refraction – change of direction of a wave on passing through an interface between two media.
resonance – the condition that the frequency of an elastic wave coincides with a natural frequency of the body being tested, or of some defect within it.
scattering – a process that changes the direction, and sometimes the type, of an elastic wave.
shear – a stress which tends to cause two parts of a body to slide over one another in a direction parallel to the plane between them.
shear wave – a wave in which the direction of particle motion is perpendicular to the direction of propagation of the wave.
Snell's law – the law relating the angles of reflection and refraction to the angle of incidence of a wave at an interface.
spectroscopy – analysis of the frequency content of signals.
specular reflection – reflection at a plane surface or near-plane segment of a surface which obeys the laws of geometrical wave propagation.
Stoneley wave – a wave confined to the interface between two solids or a fluid and a solid.
strain – a measure of the distortion of a solid material.
stress – force per unit area.
surface wave – same as Rayleigh wave *(q.v.)*.
traction – a force on a surface element.
transducer – a device for converting energy from one form to another: in this monograph, for converting electrical energy to elastic energy and *vice versa*.

References

At the end of each entry will be found the number of the first page in this monograph on which reference is made to the work in question.

Abramowitz M. and Stegun I.A. (1965) *Handbook of Mathematical Functions,* Dover, New York. 132

Achenbach J.D. (1973) *Wave Propagation in Elastic Solids,* North–Holland/American Elsevier. 4

Achenbach J.D. and Gautesen A.K. (1977) *J. Acoust. Soc. Am.* **61,** 413. 119

Achenbach J.D. and Norris A.N. (1982) *J. Nondest. Eval.* **3,** 229. 180

Achenbach J.D., Angel Y.C. and Lin W. (1984) in *Wave Propagation in Homogeneous Media and Ultrasonic Nondestructive Evaluation – AMD Volume 62,* edited by G.C. Johnson, American Society of Mechanical Engineers. 156

Achenbach J.D., Gautesen A.K. and McMaken H. (1978) in *Elastic Waves and Nondestructive Testing of Materials,* edited by Y-H. Pao, ASME New York. 124

Achenbach J.D., Gautesen A.K. and McMaken H. (1982) *Ray Methods for Waves in Elastic Solids – with Applications to Scattering by Cracks,* Pitman, London. 94

Achenbach J.D., Gautesen A.K. and Mendelson D.A. (1980) *IEEE Trans. Sonics & Ultrason.* **SU–27,** 124. 126

Achenbach J.D., Lin W. and Keer L.M. (1986) *Ultrasonics* **24,** 207. 194

Adler L. and Lewis D.K. (1976) *IEEE Trans. Sonics & Ultrason.* **SU–23,** 351. 95

Akoev G.M. (1975) *Defektoskopiya* **(1),** 65. 194

Aks S. Ø. and Vezzetti D.J. (1980) *Ultrason. Imaging* **2,** 85. 100

Alabi B. (1985) *Appl. Math. Modelling* **9,** 62. 144

Aleshin N.P. and Mogilner L.Yu. (1982) *Sov. J. Nondestr. Test.* **18,** 18. 126

Aleshin N.P. and Mogilner L.Yu. (1984) *Sov. J. Nondestr. Test.* **20,** 349. 126

Allen D.R. and Sayers C.M. (1984) *Ultrasonics* **22,** 179. 12

Allen D.R., Cooper W.H.B., Sayers C.M. and Silk M.G. (1982) in *Research Techniques in Nondestructive Testing, Volume 6,* edited by R.S. Sharpe, Academic Press. 12

Alterman Z. (1968) *J. Phys. Earth* **16,** 113. 130

Alterman Z. and Karal F.C. (1968) *Bull. Seismol. Soc. Am.* **58,** 367. 139

Alterman Z. and Loewenthal D. (1970) *Geophys. J. R. Astr. Soc.* **20,** 101. 133

Alterman Z. and Rotenberg A. (1969) *Bull. Seismol. Soc. Am.* **59,** 347. 139

Ames W.F. (1977) *Numerical Methods for Partial Differential Equations,* Academic/ Thomas Nelson. 133

Angel Y.C. and Achenbach J.D. (1984) *J. Acoust. Soc. Am.* **75,** 313. 156
Angel Y.C. and Achenbach J.D. (1985a) *J. Appl. Mech.* **107,** 33. 182
Angel Y.C. and Achenbach J.D. (1985b) *Wave Motion* **7,** 375. 182
Argyris J.H. (1960) *Energy Theorems and Structural Analysis,* Butterworths (reprinted from *Aircraft Engineering,* Oct. 1954 – May 1955). 130
Auld B.A. (1973) *Acoustic Fields and Waves in Solids,* Wiley Interscience. 4
Auld B.A. and Tan M. (1977) *Proc. 1977 Ultrason. Symp.*, 61. 169
Auld B.A. and Tsao E.M. (1977) *IEEE Trans. Sonics & Ultrason.* **SU–24,** 317. 169
Baborovsky V.M., Marsh D.M. and Slater E.A. (1973a) *Proc. Ultrasonics International Conf., 162.* 157
Baborovsky V.M., Marsh D.M. and Slater E.A. (1973b) *Non–Destr. Test.* **6,** 200. 157
Baborovsky V.M., Slater E.A. and Marsh D.M. (1975) *Proc. Ultrasonics International Conf., 46.* 157
Baerwald H. (1930) *Ann. Phys. (Germany)* **7,** 731. 18
Baik J-M. and Thompson R.B. (1984) in *Review of Progress in Quantitative Nondestructive Evaluation, Volume 4,* edited by D.O. Thompson and D.E. Chimenti, Plenum Press. 182
Baker B.B. and Copson E.T. (1950) *The Mathematical Theory of Huygens' Principle, 2nd Edition,* Oxford. 87
Banaugh R.P. and Goldsmith W. (1963) *J. Acoust. Soc. Am.* **35,** 159. 88
Barnett D.M. and Lothe J. (1974) *J. Phys. F: Metal Physics* **4,** 671. 39
Barnett D.M. and Lothe J. (1985) *Proc. R. Soc. London* **A402,** 135. 39
Barnett D.M., Lothe J., Gavazza S.D. and Musgrave M.J.P. (1985) *Proc. R. Soc. London* **A402,** 153. 40
Barnett D.M., Lothe J., Nishioka K. and Asaro R.J. (1973) *J. Phys. F: Metal Physics* **3,** 1083. 39
Basatskaya L.V. and Ermolov I.N. (1980) *Sov. J. Nondestr. Test.* **16,** 524. 59
Bass F.G. and Fuks I.M. (1979) *Wave Scattering from Statistically Rough Surfaces,* translated and edited by C.B. Vesecky and J.F. Vesecky, Pergamon. 178
Bates R.H.T. and Wall D.J.N. (1977) *Philos. Trans. R. Soc. London* **45,** 287. 110
Bayliss A. (1966) *Bull. Seismic Soc. Am.* **76,** 1115. 131
Bayliss A. and Turkel E. (1980) *Commun. Pure & Appl. Math.* **33,** 707. 144
Bažant Z.P., Glazik J.L. and Achenbach J.D. (1976) *J. Eng. Mech. Div. Proc. Am. Soc. Civil Eng.* **EM3,** 479. 131
Beaver W.L. (1974) *J. Acoust. Soc. Am.* **56,** 1043. 193
Beckmann P. and Spizzichino A. (1963) *The Scattering of Electromagnetic Waves from Rough Surfaces,* Macmillan. 94
Beltzer A.I. and Brauner N. (1985) *Appl. Phys. Lett.* **47,** 1054. 176
Berry M.V. (1976) *Adv. Phys.* **25,** 1. 124
Berry M.V. (1979) *J. Phys. A: Math. Gen.* **12,** 781. 178
Berry M.V. and Blackwell T.M. (1981) *J. Phys. A: Math. Gen.* **14,** 3101. 178
Bertholf L.D. (1967) *J. Appl. Mech.* **34,** 725. 130
Bertoni H.L. and Tamir T. (1973) *Appl. Phys.* **2,** 157. 48

Betti E. (1872) *Nuovo Cimento ser 2* **7,** 8. 60
Bhatia A.B. and Moore R.A. (1959) *J. Acoust. Soc. Am.* **31,** 1140. 175
Biot M.A. (1957) *Phys. Rev.* **105,** 1129. 18
Biot M.A. and Tolstoy I. (1957) *J. Acoust. Soc. Am.* **29,** 381. 179
Blake R.J. (1985) in *Information Quarterly for Surface Science,* informal newsletter of Daresbury Laboratory. 145
Blake R.J., Bond L.J. and Downie A.L. (1982) in *Review of Progress in Quantitative Nondestructive Evaluation Volume 1,* edited by D.O. Thompson and D.E. Chimenti, Plenum Press. 130
Bleistein N. (1966) *Commun. Pure & Appl. Math.* **19,** 353. 122
Bolomey J.C. and Wirgin A. (1974) *Proc. IEEE* **121,** 794. 112
Bond L.J. (1979) *Ultrasonics* **17,** 71. 126
Bond L.J. (1982) in *Research Techniques in Nondestructive Testing Volume 6,* edited by R.S. Sharpe, Academic Press. 130
Bond L.J. and Blake R.J. (1983) in *Proc. Ultrasonics International 83,* Butterworths. 130
Boore D.M. (1970) *Doctoral Thesis,* Massachussetts Institute of Technology. 135
Boore D.M. (1972) In *Methods in Computational Physics Volume 11,* edited by B.A. Bolt, Academic Press. 135
Bostrom A. (1980) *J. Acoust. Soc. Am.* **67,** 399. 110
Bostrom A. and Kristensson G. (1980) *Wave Motion* **2,** 335. 156
Bouwkamp C.J. (1954) *Rep. Prog. Phys.* **17,** 35. 88
Bowman J.J., Senior T.B.A. and Uslenghi P.L.E. (1969) *Electromagnetic Scattering by Simple Shapes,* North–Holland. 96
Brackbill J. (1982) in *Numerical Grid Generation,* edited by J.F. Thompson, North–Holland. 145
Brekhovskikh L.M. (1960) *Waves in Layered Media,* Academic Press, New York. 21
Brillouin L. (1914) *Ann. Phys. (France)* **44,** 203. 18
Brillouin L. (1960) *Wave Propagation and Group Veolcity,* Academic Press. 18
Brind R.J., Achenbach J.D. and Gubernatis J.E. (1984) *Wave Motion* **6,** 41. 85
Budiansky B. and Rice J.R. (1978) *J. Appl. Mech.* **45,** 453. 104
Burton A.J. and Miller G.F. (1971) *Proc. R. Soc. London* **A323,** 201. 88
Caignard L. (1962) *Reflection and Refraction of Progressive Seismic Waves,* translated and revised by E.A. Flinn and C.H. Dix, McGraw–Hill 39
Cameron R.F. and Temple J.A.G. (1985) *Int. J. Press. Vess. and Piping* **18,** 255 202
Cameron R.F. and Temple J.A.G. (1986) *Nucl. Eng. and Design* **91,** 57 202
Cauchy A. (1828) *Exercises de Math.* **3,** 328. 14
Chadwick P. and Smith G.D. (1977) *Adv. Appl. Mech.* **17, 303. 39**
Chadwick P. and Smith G.D. (1982) in *Mechanics of Solids (Rodney Hill 60th Anniversary Volume),* edited by H.G. Hopkins and M.J. Sewell, Pergamon. 39
Chapman C.H. (1978) *Geophys. J. R. Astr. Soc.* **54,** 481. 53
Chapman R.K. (1981) *Central Electricity Generating Board Report NW/SSD/RR/145/81.* 116
Chapman R.K. (1984) *Central Electricity Generating Board Report NWR/SSD/84/0059/R.* 94

Chapman R.K. and Coffey J.M. (1984) in *Review of Progress in Quantitative Nondestructive Evaluation Volume 3,* edited by D.O. Thompson and D.E. Chimenti, Plenum Press. 96

Clayton R. and Engquist B. (1977) *Bull. Seismol. Soc. Am.* **67,** 1529. 144

Clifton R.J. (1967) *Quart. Appl. Math.* **25,** 97. 131

Clifton R.J. (1985) *Appl. Mech. Rev.* **38,** 1276. 203

Clough R.W. (1960) *Proc. 2nd Conf. Electronic Computation,* ASCE Pittsburg. 130

Coffey J.M. and Chapman R.K. (1983) *Nucl. Energy* **22,** 319. 96

Coffey J.M., Chapman R.K. and Hanstock D.J. (1982) *Central Electricity Generating Board Report NWR/SSD/82/0045/R.* 201

Courant R. and Hilbert D. (1961) *Methods of Mathematical Physics,* Wiley. 127

Cruse T.A. (1968) *J. Math. Anal. & Appl.* **22,** 341. 90

Cruse T.A. and Rizzo F.J. (1968) *J. Math. Anal. & Appl.* **22,** 244. 90

Crutzen S. (1982) in *Advances in Non-destructive Examination for Structural Integrity,* Applied Science Publishers. 202

Datta S.K. (1977a) *J. Acoust. Soc. Am.* **61,** 1432. 102

Datta S.K. (1977b) *J. Appl. Mech.* **44,** 657. 177

Datta S.K. (1978) in *Mechanics Today, Volume 4,* edited by S. Nemat–Nasser, Pergamon Press. 89

Datta S.K. and Sangster J.D. (1974) *SIAM J. Appl. Math.* **26,** 350. 102

Datta S.K. and Shah A.H. (1982) *Wave Motion* **4,** 265. 131

de Billy M., Cohen–Tenoudji F., Quentin G., Lewis K. and Adler L. (1980) *J. Nondestr. Eval.* **1,** 249. 179

de Hoop A.T. (1958) *Doctoral Thesis,* Technische Hogeschool, Delft. 87

de Hoop A.T. (1985) *Wave Motion* **7,** 569. 94

Devaney A.J. (1980) *J. Math. Phys.* **21,** 2603. 177

Dewey B.R., Oliver B.F., Pickard C.A. and Adler L. (1983) in *Review of Progress in Quantitative Nondestructive Evaluation Volume 2,* edited by D.O. Thompson and D.E. Chimenti, Plenum Press. 131

Dick P. (1977) *Mater. Eval.* **35,** 26. 2

Dickey J.W., Frisk G.V. and Überall H. (1976) *J. Acoust. Soc. Am.* **59,** 1339. 85

Domany E. and Entin–Wohlman O. (1984) *J. Appl. Phys.* **56,** 132. 110

Domany E., Entin–Wohlman O. and Mizrachi L. (1984) *J. Appl. Phys.* **56,** 132. 110

Domany E., Teitel S. and Krumhansl J.A. (1978) *J. Appl. Phys.* **49,** 2599. 104

Doolittle R.D., Überall H. and Ugincius P. (1968) *J. Acoust. Soc. Am.* **43,** 1. 85

Doyle P.A. (1980) *J. Phys. D: Appl. Phys.* **13,** 163. 123

Doyle P.A., Latimer J. and Adler L. (1982) in *Research Techniques in Nondestructive Testing Volume 5,* edited by R.S. Sharpe, Academic Press. 123

Drury J.C. (1978) *Ultrasonic Flaw Detection for Technicians,* Unit Inspection Company, British Steel Corporation. 4

Duncan D.B. (1985) in *Proc. 5th Canadian Conf. Nondest. Test.* 131

Einspruch N.G. and Truell R. (1960) *J. Acoust. Soc. Am.* **32,** 214. 71

Einspruch N.G., Witterholt E.J. and Truell R. (1960) *J. Appl. Phys.* **31,** 806. 71

Elices M. and Garcia–Moliner F. (1968) in *Physical Acoustics, Volume 5,* edited by W.P. Mason, Academic Press. 18

Engquist B. and Majda A. (1977) *Math. Comput.* **31,** 629. 144
Engquist B. and Majda A. (1979) *Commun. Pure & Appl. Math.* **32,** 313. 144
Eshelby J.D. (1957) *Proc. R. Soc. London* **A271,** 376. 103
Euler L. (1768) *Instructiones Calculi Integralis,* St. Petersburg 130
Evans A.G., Tittmann B.R., Ahlberg L., Khuri–Yakub B.T. and Kino G.S. (1978) *J. Appl. Phys.* **49,** 2669 176
Ewing M., Jardetzky W. and Press F. (1957) *Elastic Waves in Layered Media,* McGraw–Hill. 50
Fedorov F.I. (1968) *Theory of Elastic Waves in Crystals,* translated by J.E.S. Bradley, Plenum Press. 10
Floyd E.R. (1984) *J. Acoust. Soc. Am.* **75,** 803. 124
Floyd E.R. (1986) *J. Acoust. Soc. Am.* **79,** 1741. 124
Fokkema J.T. (1980) *Wave Motion* **2,** 375. 178
Fokkema J.T. (1981) *Wave Motion* **3,** 33. 178
Fokkema J.T. and van den Berg P.M. (1977) *J. Acoust. Soc. Am.* **62,** 1095. 178
Fokkema J.T. and van den Berg P.M. (1980) *J. Acoust. Soc. Am.* **68,** 1836. 178
Folds D.L. and Loggins C.D. (1977) *J. Acoust. Soc. Am.* **62** 1102. 21
Foldy L.L. (1945) *Phys. Rev.* **67,** 107. 177
Forsyth A.R. (1927) *Calculus of Variations,* Cambridge. 127
Foster F.S. and Hunt J.W. (1978) *Ultrasonics* **16,** 116. 193
Freedman A. (1962) *Acustica* **12,** 10. 95
Friedlander F.G. (1948) *Q. J. Mech. & Appl. Math.* **1,** 376. 52
Friedlander F.G. (1958) *Sound Pulses,* Cambridge University Press. 113
Friedman M.B. and Shaw R. (1962) *J. Appl. Mech.* **29,** 40. 98
Frielinghaus R., Krautkrämer J. and Schlengermann U. (1970) *Non–Destr. Test.* **3,** 125. 161
Frisk G.V., Dickey J.W. and Überall H. (1975) *J. Acoust. Soc. Am.* **58,** 996. 37
Gautesen A.K. (1980) *Wave Motion* **1,** 127. 122
Gautesen A.K., Achenbach J.D. and McMaken H. (1978) *J. Acoust. Soc. Am.* **63,** 1824. 121
Georgiou G.A. and Bond L.J. (1985) in *Proc. 4th Int. Symp. Offshore Mech. and Arctic Engineering.* 144
Gericke O.R. (1963) *J. Acoust. Soc. Am.* **35,** 364. 96
Gericke O.R. (1966) *J. Met.* **18,** 932. 96
Gilbert F. and Laster S.J. (1962) *Bull. Seismol. Soc. Am.* **52,** 299. 37
Glauber R.J. (1959) in *Lectures in Theoretical Physics,* edited by W.E. Brittin and L.G. Dunham, Interscience. 113
Gniadek K. and Petykiewicz J. (1971) in *Progress in Optics, Volume 9,* edited by E. Wolf, North–Holland. 87
Goebbels K. (1975) *Materialprüf.* **17,** 231. 178
Goebbels K. (1976) *Materialprüf.* **18,** 86. 178
Goebbels K. and Höller P. (1976) in *Proc. 8th World Conf. NDT, Cannes, France.* 178
Goebbels K. and Höller P. (1978) in *Proc. 1st Intern. Symp. Materials Charact., Gaithersburg, USA.* 178
Golan S., Adler L., Cook K.V., Nanstad R.K. and Bolland T.K. (1980) J. Nondestr. Eval. **1,** 11. 155
Goos F. and Hänchen H. (1947) *Ann. Phys. (Germany)* **1,** 333. 48
Graff K.F. (1975) *Wave Motion in Elastic Solids,* Oxford University Press. 4
Graff K.F. (1981) in *Physical Acoustics Vol XV,* edited by W.P. Mason and R.N. Thurston, Academic Press. 3

Green A.T., Lockman C.S. and Steel R.K. (1964) *Mod. Plastics* **41,** 137. 54

Green G. (1839) *Trans. Cambridge Philos. Soc.* **7,** 121. 9

Greenwood J.A. (1984) *Proc. R. Soc. London* **A393,** 133. 178

Gubernatis J.E. (1979) *J. Appl. Phys.* **50,** 4046. 101

Gubernatis J.E., Domany E. and Krumhansl J.A. (1977) *J. Appl. Phys.* **48,** 2804. 90

Gubernatis J.E., Domany E. and Krumhansl J.A. (1979) *Los Alamos Scientific Laboratory Report LA-UR–2393.* 125

Gubernatis J.E., Domany E., Krumhansl J.A. and Huberman M. (1977) *J. Appl. Phys.* **48,** 2812. 71

Gubernatis J.E., Krumhansl J.A. and Thomson R.M. (1979) *J. Appl. Phys.* **50,** 3338. 104

Gunton D.J. and Saunders G.A. (1975) *Proc. R. Soc.London* **A343,** 63. 11

Gupta N. and Kisslinger C. (1966) *Geophysics* **31,** 1057. 153

Hackman R.H. and Todroff D.G. (1985) *J. Acoust. Soc. Am.* **78,** 1058. 112

Haines N.F. (1980) *Central Electricity Generating Board Report RD/B/N4744. 180*

Haines N.F. and Langston D.B. (1980) *J. Acoust. Soc. Am.* **67,** 1443. 88

Hall K.G. (1976) unpublished work. See photographs in article by M.G. Silk in *Research Techniques in Nondestructive Testing, Volume 3,* edited by R.S. Sharpe, Academic Press. 157

Hall K.G. (1977) *Ultrasonics* **15,** 245. 157

Harker A.H. (1985) *J. Nondestr. Eval.* **4,** 89. 170

Hart R.W. and Farrell R.A. (1977) *IEEE Trans. Ant. Propag.* **AP–25,** 708. 179

Harumi K. (1961) *J. Appl. Phys.* **32,** 1488. 146

Harumi K. (1962) *J. Appl. Phys.* **33,** 3588. 146

Harumi K. (1986) *NDT International* **19,** 315. 130

Harumi K., Igarashi T. and Saito T. (1978) *J. Japan. NDT Soc.* **27,** 807. 132

Harumi K., Saito T. and Fujimori T. (1979) in *Proc. 9th World Conf. NDT, Melbourne, Australia. 130*

Harumi K., Suzuki F. and Saito T. (1973) *J. Acoust. Soc. Am.* **53,** 600. 130

Hashin Z. and Shtrikman S. (1962a) *J. Mech. Phys. Solids* **10** 335. 129

Hashin Z. and Shtrikman S. (1962b) *J. Mech. Phys. Solids* **10** 343. 129

Hashin Z. and Shtrikman S. (1963) *J. Mech. Phys. Solids* **11** 127. 129

Hayman A.J. (1977) *Schlieren visualization of focused ultrasonic images,* Ph. D. Thesis, City University, London. 160

Heaps H.S. (1957) *J. Appl. Phys.* **28,** 815. 179

Heyman E. and Felsen L.B. (1985) *Wave Motion* **7,** 335. 53

Hickling, R. (1962) *J. Acoust. Soc. Am.* **34,** 1582. 164

Hirsekorn S. (1982) *J. Acoust. Soc. Am.* **72,** 1021. 175

Hudson J.A. (1968) *Q. J. Mech. & Appl. Math.* **21,** 487. 175

Hudson J.A. (1980) *The Excitation and Propagation of Elastic Waves,* Cambridge University Press. 7

Hulthén L. (1944) *Kgl. Fysiograf. Sällskap Lund Förh.* **14**(21). 127

Hulthén L. (1947) *Comptes Rend. Dixième Congr. Mat. Scandinaves,* page 201. 127

Hulthén L. (1948) *Arkiv Mat. Astron. Fys.* **35A M(25). 127**

Huntington H. (1958) *Solid State Physics* **7,** 213. 9

Hyman J.M. (1982) in *Numerical Grid Generation,* edited by J.F. Thompson, North–Holland. 145

Ilan A. and Bond L.J. (1979) *Geophys. J. R. Astr. Soc.* **65,** 75. 160

Ilan A. and Loewenthal D. (1976) *Geophys. Prospect.* **24,** 431. 134

Ilan A., Bond L.J. and Spivack M. (1979) *Geophys. J. R. Astr. Soc.* **57,** 463. 160

Ilan A., Ungar A. and Alterman Z. (1975) *Geophys. J. R. Astr. Soc.* **43,** 727. 139

Johnson D.M. (1976) *J. Acoust. Soc. Am.* **59,** 1319. 88

Johnson J. (1985) *Mater. Eval.* **43,** 1489. 197

Kaiser J. (1953) *Arch. Eisenhüttenwes.* **24,** 43. 54

Karal F.C. and Keller J.B. (1959) *J. Acoust. Soc. Am.* **31,** 694. 113

Kausel E. and Tassoulas J. (1981) *Bull. Seismol. Soc. Am.* **71,** 143. 144

Keller J.B. (1953) *Proc Symposium on Microwave Optics,* McGill University. 113

Keller J.B. (1957) *J. Appl. Phys.* **28,** 426–444. 113

Keller J.B. (1979) *SIAM Rev.* **21,** 229. 113

Kendall K. and Tabor D. (1971) *Proc. R. Soc. London* **A 323,** 321. 180

Keogh P.S. (1983) *Doctoral Thesis,* University of Manchester. 96

Keogh P.S. (1986) *Q. J. Mech. & Appl. Math.* **39,** 535. 126

Kino G.S. (1978) *J. Appl. Phys.* **49,** 3190. 66

Kirby C. (1984) *Water in Great Britain,* Penguin. 2

Kittinger E., Tichý J. and Bertagnolli E. (1981) *Phys. Rev. Lett.* **47,** 712. 11

Kline M.S. and Kay I.W. (1965) *Electromagnetic Theory and Geometrical Optics,* Interscience. 113

Knopoff L. (1956) *J. Acoust. Soc. Am.* **28,** 217. 87

Knopoff L. (1959a) *Geophysics* **24,** 30. 71

Knopoff L. (1959b) *Geophysics* **24,** 209. 71

Knops R.J. and Payne L.E. (1971) *Uniqueness Theorems in Linear Elasticity,* Springer–Verlag. 12

Knott J.F. (1973) *Fundamentals of Fracture Mechanics,* Butterworths. 186

Ko W.L. and Karlsson T. (1967) *J. Appl. Mech.* **34,** 921. 98

Kobayashi S. and Nishimura N. (1982) in *Developments in Boundary Element Methods – 2,* edited by P.K. Banerjee and R.P. Shaw, Applied Science Publishers. 90

Korringa J. (1979) in *Acoustic, Electromagnetic and Elastic Wave Scattering – Focus on the t-matrix Approach,* edited by V.K. Varadan and V.V. Varadan, Pergamon Press. 177

Koshiba M., Karakida S. and Suzuki M. (1984) *IEEE Trans. Sonics & Ultrason.* **SU–31,** 18. 169

Koshiba M., Karakida S., Morita H. and Suzuki M. (1983) *Jpn. J. Appl. Phys.* **22, suppl 3,** 148. 131

Kraut E.A. (1976) *IEEE Trans. Sonics & Ultrason.* **SU–23,** 162. 89

Krautkrämer J. and Krautkrämer H. (1982) *Ultrasonic Testing of Materials,* Springer–Verlag. 4

Kronig R. and Kramers H.A. (1928) *Z. Phys.* **48,** 174. 19

Kupradze V.D. (1963) in *Dynamical Problems in Elasticity, Progress in Solid Mechanics Vol III,* edited by I.N. Sneddon and R. Hill, North–Holland. 60

Lakes R (1987) *Science* **235,** 1038 11

Lam F.K. and Tsang W.M. (1985) *Ultrasonics* **23,** 14. 95

Lamb H. (1904) *Phil. Trans. R. Soc. London* **A203,** 1. 53

Lamé G. (1852) *Leçons sur la théorie mathématique de l'élasticité des corps solides,* Paris. 10

Landau L.D. and Lifshitz E.M. (1958) *Statistical Physics,* Pergamon Press. 19
Langenberg K.J. and Schmitz V. (1986) *Nucl. Eng. and Design* **94,** 427. 126
Lax M. (1951) *Rev. Mod. Phys.* **23,** 287. 177
Lax M. (1952) *Phys. Rev.* **85,** 621. 177
Lax P.D. and Wendroff B. (1960) *Commun. Pure & Appl. Math.* **13,** 217. 139
Lax P.D. and Wendroff B. (1964) *Commun. Pure & Appl. Math.* **17,** 381. 139
Lee D.A. and Corbly D.M. (1977) *IEEE Trans. Sonics & Ultrason.* **SU–24,** 206. 40
Levine H. and Schwinger J. (1948) *Phys. Rev.* **74,** 958. 127
Levine H. and Schwinger J. (1949) *Phys. Rev.* **75,** 1423. 127
Lewis R.M. and Boersma J. (1969) *J. Math. Phys.* **10,** 2291. 122
Lewis T.S., Kraft D.W. and Hom N. (1976) *J. Appl. Phys.* **47,** 1795. 80
Lidington B.H., Saunderson D.H. and Silk M.G. (1974) *United Kingdom Atomic Energy Authority Report AERE-R7830.* 161
Lifshitz I.M. and Parkomovsky G.D. (1950) *Zh. Eksp. & Teor. Fiz.* **20,** 175. 174
Lighthill M.J. (1965) *J. Inst. Math. & Appl.* **1,** 1. 18
Lighthill M.J. (1978) *Waves in Fluids,* Cambridge University Press. 113
Lindman E.L. (1975) *J. Comput. Phys.* **18,** 66. 144
Liska R. (1984) *Comput. Phys. Commun.* **34,** 175. 136
Liszka E.G. and McCoy J.J. (1982) *J. Acoust. Soc. Am.* **71,** 1093. 179
Lothe J. and Barnett D.M. (1976) *J. Appl. Phys.* **47,** 428. 39
Love A.E.H. (1911) *Some Problems of Geodynamics,* Cambridge University Press. 39
Love A.E.H. (1944) *A Treatise on the Mathematical Theory of Elasticity,* Dover. 11
Lumb R.F. (1977) *Phys. in Technol.* **8,** 249. 161
Lumb R.F., Bosselaar H. and Baborovsky V.M. (1978) *Mater. Eval.* **36,** No. 10, page 57. 161
Luneberg R.K. (1964) *Mathematical Theory of Optics,* University of California Press. 113
Lynch P.J. (1970) *J. Acoust. Soc. Am.* **47,** 804. 179
Lysmer J. and Drake L.A. (1972) in *Methods in Computational Physics Vol 11,* edited by B.A. Bolt, Academic Press. 131
Lysmer J. and Kuhlemeyer R.L. (1967) *J. Eng. Mech. Div. Proc. Am. Soc. Civil Eng.* **95,** 859. 144
Mal A.K. (1968) *Q. J. Appl. Math.* **26,** 231. 89
Mal A.K. (1970) *Int. J. Eng. Sci.* **8,** 381. 89
Mandelbrot B.B. (1977) *Fractals, Form, Chance and Dimension,* Freeman. 178
Mandelbrot B.B. (1982) *The Fractal Geometry of Nature,* Freeman. 178
Mandelbrot B.B., Passoja D.E. and Paullay A.J. (1984) *Nature* **308,** 721. 178
Manolis G.D. (1983) *Int. J. Numer. Meth. Eng.* **9,** 73. 90
Manolis G.D. and Beskos D.E. (1980) *Proc. 2nd Int. Symp. Innovative Numer. Anal. Appl. Eng. Sci.,* 459. 90
Manolis G.D. and Beskos D.E. (1981) *Int. J. Numer. Meth. Eng.* **17,** 573. 90
Maradudin A.A. and Mills D.L. (1976) *Ann. Phys. (USA)* **100,** 262. 179
Marsh K.A., Richardson J.M., Addison R.C. and Elsley R.K.(1982) in *Review of Progress in Quantitative Nondestructive Evaluation*

Volume 2, edited by D.O. Thompson and D.E. Chimenti, Plenum Press. 202
Marshall W. (1976) *An assessment of the integrity of PWR pressure vessels,* report of a UKAEA study group. 202
Marshall W. (1982) *An assessment of the integrity of PWR pressure vessels,* second report of a UKAEA study group. 202
Martin P.A. (1981) *Proc. R. Soc. London* **A378,** 263. 89
Martin P.A. and Wickham G.R. (1983) *Proc. Roy. Soc.* **A390,** 91. 89
Mason W.P. and McSkimin H.J. (1948) *J. Appl. Phys.* **19,** 940. 176
Maue A.W. (1953) *Z. Angew. Math. & Mech.* **33,** 1. 116
Maxey W.A. (1974) *Proc. Fifth Symp. Linepipe Research,* American Gas Association. 186
McClure G.M., Duffy A.R. and Eiber R.J. (1965) *J. Eng. Ind.* **87,** 265. 186
McMaken H. (1984) *J. Acoust. Soc. Am.* **75,** 1352. 124
Medwin H. and Novarini J.C. (1981) *J. Acoust. Soc. Am.* **69,** 108. 179
Meecham W.C. (1956) *J. Appl. Phys.* **27,** 361. 179
Meecham W.C. (1956) *J. Rat. Mech. Anal.* **5,** 2. 95
Mendelson D.A., Achenbach J.D. and Keer L.M. (1980) *Wave Motion* **2,** 277. 126
Merkulov L.G. and Rokhlin S.I. (1970) *Defektoskopiya* **3,** 13 (in Russian). 169
Millar R.F. (1973) *Radio Sci.* **8,** 785. 110
Miller G.F. and Pursey H. (1954) *Proc. R. Soc. London* **A223,** 521. 54
Miller G.F. and Pursey H. (1955) *Proc. R. Soc. London* **A233,** 55. 54
Mills G.S. (1974) *Mater. Eval.* **32,** 256. 96
Morette C. (1951) *Phys. Rev.* **81,** 848. 113
Morse P.M. and Feshbach H. (1953) *Methods of Theoretical Physics,* McGraw Hill. 60
Morton K.W. (1976) *Comput. Phys. Commun.* **12,** 99. 130
Mühlhäuser O. (1931) German patent *Verfahren zu Zustandsbestimmung von Werkstoffen, besonders zur Ermittlung von Fehlern darin,* DRP Nr 569598. 3
Munasinghe M. and Farnell G.W. (1973) *J. Geophys. Res.* **78,** 2454. 146
Musgrave M.J.P. (1978) in *Modern Problems in Elastic Wave Propagation,* edited by J. Miklowitz and J.D. Achenbach, Wiley–Interscience. 17
Neubauer W.G. (1963) *J. Acoust. Soc. Am.* **35,** 279. 88
Newton R.G. (1966) *Scattering Theory of Waves and Particles,* McGraw Hill. 106
Niwa S., Hirose S. and Kitahara M. (1986) *Wave Motion* **8,** 77. 90
Niwa Y., Kitihara M. and Ikeda H. (1984) in *Theoretical and Applied Mechanics* **32, University of Tokyo Press, page 183. 90**
Novarini J.C. and Medwin H. (1978) *J. Acoust. Soc. Am.* **64,** 260. 179
Nye J.F. (1957) *Physical Properties of Crystals,* Oxford University Press. 7
O.G.J. (1960) *Oil & Gas J.* **58** (17), 105. 186
O'Brien C.G., Hyman M.A. and Kaplan S. (1951) *J. Math. Phys.* **29,** 223. 133
O'Donnell M., Jaynes E.T. and Miller J.G. (1981) *J. Acoust. Soc. Am.* **69,** 696. 20
Ogilvy J.A. (1985) *NDT Int.* **18,** 67. 176
Ogilvy J.A. (1986) *J. Phys. D: Applied Phys.* **19,** 2085. 129
Ogilvy J.A. (1987) *Ph. D. Thesis,* University of Bath. 88
Ogilvy J.A. and Temple J.A.G. (1983) *Ultrasonics* **21,** 259. 116

Ootsuki A. and Harumi K. (1983) *Eng. Struct. Dynamics* **11,** 441. 131
Opsal J.L. and Visscher W.M. (1985) *J. Appl. Phys.* **58,** 1102. 111
Pao Y-H. (1978) in *Modern Problems in Elastic Wave Propagation,* edited by J. Miklowitz and J.D. Achenbach, Wiley–Interscience. 110
Pao Y-H. and Mow C.C. (1973) *Diffraction of Elastic Waves and Dynamic Stress Concentrations,* Adam Hilger/Crane Russak. 71
Pao Y-H. and Mow C.C. (1976) *J. Acoust. Soc. Am.* **59,** 1046. 83
Pao Y-H. and Sachse W. (1974) *J. Acoust. Soc. Am.* **56,** 1478. 167
Pao Y-H. and Varatharajulu V. (1976) *J. Acoust. Soc. Am.* **59,** 1361. 60
Pao Y-H., Sachse W. and Fukuoka H. (1984) in *Physical Acoustics, Volume 17,* edited by W.P. Mason and R.N. Thurston, Academic Press. 12
Papadakis E.P. (1965) *J. Acoust. Soc. Am.* **37,** 703. 175
Papadakis E.P. (1968) in *Physical Acoustics, Volume 4B,* edited by W.P. Mason, Academic Press. 175
Papadakis E.P. (1984) *Int. Met. Rev.* **29,** 1. 176
Peterson B. and Strom S. (1973) *Phys. Rev.* **D8,** 3661. 110
Peterson B. and Strom S. (1974a) *J. Acoust. Soc. Am.* **56,** 771. 110
Peterson B. and Strom S. (1974b) *Phys. Rev.* **D10,** 2670. 110
Peterson B. and Strom S. (1975) *J. Acoust. Soc. Am.* **57,** 2. 110
Pitts L.E. and Mayer W.G. (1977) *Ultrasonics* **15,** 201. 48
Pitts L.E., Plona T.J. and Mayer W.G. (1977) *IEEE Trans. Sonics & Ultrason.* **SU24,** 101. 48
Poisson S.D. (1829) *Mem. de l'Acad. Royale des Sci.* **8,** 357. 14
Poston T. and Stewart I.N. (1978) *Catastrophe Theory and its Applications,* Pitman. 124
Poynton W.A. (1974) *Proc. I.G.E. Symp. Crack Propagation in Pipelines, Newcastle-upon–Tyne.* 186
Pritchard S.E. (1987) *NDT International* **20,** 57. 12
Punjani M. and Bond L.J. (1986) in *Review of Progress in Quantitative Nondestructive Evaluation Volume 5,* edited by D.O. Thompson and D.E. Chimenti, Plenum Press. 182
Rayleigh, Lord (1873) *Proc. London Math. Soc.* **4,** 357. 60
Rayleigh, Lord (1885) *Proc. London Math. Soc.* **17,** 4. 34
Rayleigh, Lord (1910) *Philos. Mag.* **20,** 1001. 83
Reut Z. (1982) in *University of Strathclyde Seminars in Applied Mathematical Analysis: Vibration Theory,* edited by G.F. Roach, *Shiva Mathematics Series* **4,** 105. 88
Reut Z. (1985) *J. Sound and Vib.* **103,** 297 88
Reynolds W.N. and Smith R.L. (1984) *J. Phys.* **D 17,** 109. 176
Rice S.O. (1944) *Bell Syst. Tech. J.* **23,** 282. 178
Rice S.O. (1945) *Bell Syst. Tech. J.* **24,** 46. 178
Richardson J.M. (1979) *Int. J. Eng. Sci.* **17,** 83. 180
Richardson J.M. (1982) in *Review of Progress in Quantitative Nondestructive Evaluation Volume 2,* edited by D.O. Thompson and D.E. Chimenti, Plenum Press. 202
Richardson J.M. (1984) *Wave Motion* **6,** 325. 106
Roberts R.A., Brind R.J. and Achenbach J.D. (1981) *Wave Motion* **2,** 153. 178
Rokhlin L.L. (1972) *Sov. Phys. Acoust.* **18,** 71. 176
Rokhlin S.I. (1980a) *J. Acoust. Soc. Am.* **67,** 1157. 169
Rokhlin S.I. (1980b) *Proc. 1980 Ultrason. Symp.,* 922. 169
Rokhlin S.I., Hefets M. and Rosen M. (1980) *J. Appl. Phys.* **52,** 3579. 40
Rose J.L. (1975) *Br. J. Non–Destr. Test.* **17,** 109. 100

Rose J.L. and Meyer P.A. (1975) *J. Acoust. Soc. Am.* **57,** 598. 130
Rulf B. (1967) *J. Math. Phys.* **8,** 1785. 84
Rulf B. (1969) *J. Acoust. Soc. Am.* **45,** 493. 85
Runge C. (1908) *Z. Math. Phys.* **56,** 225. 130
Sachse W., Hsu N.N. and Eitzen D.G. (1978) *Proc. IEE Symposium,* 139. 160
Saito T., Fujimori T., Harumi K. and Okada H. (1982) in *Proc. 10th World Conf. NDT, Moscow. 130*
Saltzman J. and Brackbill J. (1982) in *Numerical Grid Generation,* edited by J.F. Thompson, North–Holland. 145
Sanford R.J. and Beanbein L.A. (1977) *Exp. Mech.* **17,** 441. 157
Sayers C.M. (1980) *J. Phys. D: Appl. Phys.* **13,** 179. 177
Sayers C.M. (1981) *J. Phys. D: Appl. Phys.* **14,** 413. 177
Sayers C.M. (1982) *J. Phys. D: Appl. Phys.* **15,** 2157. 12
Sayers C.M. and Smith R.L. (1982) *Ultrasonics* **20,** 201. 177
Scandrett C.L. and Achenbach J.D. (1987) *Wave Motion* **9,** 171. 130
Scandrett C.L., Kriegsmann G.A. and Achenbach J.D. (1986a) *J. Comput. Phys.* **65,** 410. 130
Scandrett C.L., Kriegsmann G.A. and Achenbach J.D. (1986b) *SIAM J. Sci. Stat. Comput.* **7,** 571. 130
Schenck H.A. (1968) *J. Acoust. Soc. Am.* **44,** 41. 88
Schoch A. (1950) *Ergeb. Exakt. Naturw.* **23,** 127. 48
Schwinger J. (1947) *Phys. Rev.* **72,** 742. 127
Schwinger J. (1950) *Phys. Rev.* **78,** 135. 127
Scruby C.B. and Wadley H.N.G. (1983) *Prog. Nucl. Energy* **11,** 275. 54
Scruby C.B., Wadley H.N.G., Dewhurst R.J., Hutchins D.A. and Palmer S.B. (1981) *Mater. Eval.* **39,** 1250. 150
Sharpe R.S. (ed.) (1970) *Research Techniques in NDT, Volume 1,* Academic Press. 4
Sharpe R.S. (ed.) (1973) *Research Techniques in NDT, Volume 2,* Academic Press. 4
Sharpe R.S. (ed.) (1977) *Research Techniques in NDT, Volume 3,* Academic Press. 4
Sharpe R.S. (ed.) (1980) *Research Techniques in NDT, Volume 4,* Academic Press. 4
Sharpe R.S. (ed.) (1982) *Research Techniques in NDT, Volume 5,* Academic Press. 4
Sharpe R.S. (ed.) (1982) *Research Techniques in NDT, Volume 6,* Academic Press. 4
Sharpe R.S. (ed.) (1984) *Research Techniques in NDT, Volume 7,* Academic Press. 4
Sharpe R.S. (ed.) (1985) *Research Techniques in NDT, Volume 8,* Academic Press. 4
Shkarlet Y.M. (1974) *Sov. J. Nondestr. Test.* **10,** 5. 169
Sih G.C. and Loeber J.F. (1969) *J. Acoust. Soc. Am.* **46,** 193. 89
Silk M.G. (1977) in *Research Techniques in Nondestructive Testing, Volume 3,* edited by R.S. Sharpe, Academic Press. 154
Silk M.G. (1984) *Ultrasonic Transducers for Nondestructive Testing,* Adam Hilger. 4
Silk M.G., Stoneham A.M. and Temple J.A.G. (1987) *The Reliability of Non-destructive Testing,* Adam Hilger. 202
Smith E.D., Richardson I. and van der Merwe G.J.J. (1987) *Br. J. Non–Destr. Test.* **29,** 157. 12
Smith W.D. (1974) *J. Comput. Phys.* **15,** 492. 144
Smith W.D. (1975) *Geophys. J. R. Astr. Soc.* **42,** 747. 131

Sokolnikov I.S. (1956) *Mathematical Theory of Elasticity*, McGraw–Hill. 10
Sokolov S.Ya. (1929) *Elek. Nachr. Tech.* **6,** 454. 3
Sternberg E. (1960) *Arch. Ration. Mech. & and Anal. (Germany)* **6,** 34. 14
Stokes G.G. (1876) *Smith's Prize examination question 11,* reprinted in *Mathematics and Physics Papers* **5,** 362 (1905), Cambridge University Press 17
Stoneley R. (1924) *Proc. R. Soc. London* **A 106,** 416. 39
Stroh A.N. (1962) *J. Math. Phys.* **41,** 77. 16
Szilard J. (ed) (1982) *Ultrasonic Testing,* John Wiley. 4
Tan M. and Auld B.A. (1980) *Proc. 1980 Ultrason. Symp.*, 857. 169
Tan T.H. (1975) *Appl. Sci. Res.* **31,** 363. 62, 89
Tan T.H. (1976) *Appl. Sci. Res.* **32,** 97. 89
Tan T.H. (1977a) *Appl. Sci. Res.* **33,** 75. 89
Tan T.H. (1977b) *Appl. Sci. Res.* **33,** 89. 89
Tan T.H. (1979) *J. Acoust. Soc. Am.* **59,** 1265. 94
Taylor J.R. (1972) *Scattering Theory,* Wiley. 106
Teitel S. (1978) *J. Appl. Phys.* **49,** 5763. 104
Temple J.A.G. (1985) *Int. J. Pres. Ves. & Piping* **19,** 185. 180
Temple J.A.G., Silk M.G. and Lidington B.H. (1979) unpublished work. 154
Thom R. (1975) *Structural Stability and Morphogenesis,* Benjamin. 124
Thomas T.R. (editor) (1982) *Rough Surfaces,* Longman. 178
Thompson J.F. (1982) in *Numerical Grid Generation,* edited by J.F. Thompson, North–Holland. 145
Thompson J.F., Warsi Z.U.A. and Mastin C.W. (1982) *J. Comput. Phys.* **47,** 1. 145
Thompson R.B. and Fiedler C.J. (1983) in *Review of Progress in Quantitative Nondestructive Evaluation, Volume 3,* edited by D.O. Thompson and D.E. Chimenti, Plenum Press. 182
Thompson R.B. and Thompson D.O. (1985) *Proc. IEEE* **73,** 1716. 4
Thomson W.T. (1950) *J. Appl. Phys.* **21** 89. 21
Tiratsoo J.N.H. (1983) *World Pipelines,* Pipes and Pipelines International. 2
Titchmarsh E.C. (1948) *Introduction to the Theory of Fourier Integrals,* Oxford University Press. 62
Tobocman W. (1984a) *J. Acoust. Soc. Am.* **76,** 599. 88
Tobocman W. (1984b) *J. Acoust. Soc. Am.* **76,** 1549. 88
Tobocman W. (1985) *J. Acoust. Soc. Am.* **77,** 369. 112
Tolstoy I. (1973) *Wave Propagation,* McGraw–Hill. 51
Trorey A.W. (1970) *Geophysics* **35,** 762. 94
Tsao S-J., Varadan V.V. and Varadan V.K. (1982) in *Review of Progress in Quantitative Nondestructive Evaluation – 2A,* edited by D.O. Thompson and D.E. Chimenti, Plenum Press. 90
Tsao S-J., Varadan V.V. and Varadan V.K. (1983) *J. Appl. Mech.* **50,** 143. 131
Turner M.J., Clough R.W., Martin H.C. and Topp L.J. (1956) *J. Aeronaut. Sci.* **23,** 805, 854. 130
Überall H. (1973) in *Physical Acoustics Volume 10,* edited by W.P. Mason and R.N. Thurston, Academic Press. 84
Überall H. (1978) in *Modern Problems in Elastic Wave Propagation,* edited by J. Miklowitz and J.D. Achenbach, Wiley–Interscience. 83
van der Hijden J.H.M.T and Neerhoff F.L. (1984) *J. Acoust. Soc. Am.* **75,** 1694. 156

Varadan V.K., Varadan V.V. and Pao Y-H. (1978) *J. Acoust. Soc. Am.* **63,** 1310. 177
Varadan V.V. and Varadan V.K. (1981) *J. Acoust. Soc. Am.* **70,** 213. 110
Varadan V.V. and Varadan V.K. (1982) in *Elastic Wave Scattering and Propagation,* Ann Arbor Science Publishers. 156
Varadan V.V., Pillai T.A.K. and Varadan V.K. (1983) *J. Appl. Mech.* **50,** 802. 156
Varadan V.V., Tsao S-J. and Varadan V.K. (1984) in *Review of Progress in Quantitative Nondestructive Evaluation Volume 3,* edited by D.O. Thompson and D.E. Chimenti, Plenum Press. 110
Varadan V.V., Wall D.J.N., Tsao S-J. and Varadan V.K. (1982) *J. Acoust. Soc. Am.* **71,** 1384. 110
Varatharajulu V. and Pao Y-H. (1976) *J. Acoust. Soc. Am.* **60,** 556 107
Viktorov I.A. (1967) *Rayleigh and Lamb Waves,* Plenum Press. 37
Visscher W.M. (1980a) *J. Appl. Phys.* **51,** 825. 110
Visscher W.M. (1980b) *J. Appl. Phys.* **51,** 835. 110
Visscher W.M. (1981) *Wave Motion* **3,** 49. 112
Visscher W.M. (1984) *J. Apply. Phys.* **56,** 713. 156
Visscher W.M. (1985) *J. Apply. Phys.* **57,** 1538. 156
Vopilkin A.Kh. (1985a) *Sov. J. Nondestr. Test.* **21,** 18. 126
Vopilkin A.Kh. (1985b) *Sov. J. Nondestr. Test.* **21,** 143. 126
Waterman P.C. (1969) *J. Acoust. Soc. Am.* **45,** 1417. 110
Waterman P.C. (1971) *Phys. Rev.* **D3,** 825. 110
Waterman P.C. (1976) *J. Acoust. Soc. Am.* **60,** 567 107
Waterman P.C. (1980) in *Acoustic, Electromagnetic and Elastic Wave Scattering – Focus on the T-Matrix Approach,* edited by V.K. Varadan and V.V. Varadan, Pergamon Press. 111
Waterman P.C. and Truell R. (1961) *J. Math. Phys.* **2,** 512. 177
Weaver R.L. (1985) *Wave Motion* **7,** 105. 128
Weight J.P. (1982) *The propagation and reception of wide-band ultrasonic pulses,* Ph. D. Thesis, City University, London. 160
Weight J.P. (1987) *J. Acoust. Soc. Am.* **81,** 815. 160
Werneyer R. and Schlengermann U. (1971) *Materialprüfung* **13,** 213. 161
Whapham A.D., Perring S. and Rusbridge K.L. (1985) *United Kingdom Atomic Energy Authority Report AERE R10854.* 182
Wheeler L.T. and Sternberg E. (1968) *Arch. Ration. Mech. & Anal. (Germany)* **31,** 51. 12
White R.M. (1958) *J. Acoust. Soc. Am.* **30,** 771. 80
Whitham B.J. (1974) *Linear and Non-linear Waves,* Wiley. 113
Williams K.L. and Marston P.L. (1985) *J. Acoust. Soc. Am.* **78,** 1093. 85
Williams K.L. and Marston P.L. (1986) *J. Acoust. Soc. Am.* **79,** 1702. 85
Williams S. and Mudge P.J. (1985) *NDT International* **18,** 123. 202
Willis J.R. (1981) *Wave Motion* **3,** 1. 177
Ying C.F. and Truell R. (1956) *J. Appl. Phys.* **27,** 1086. 71
Ying C.F., Li M.X. and Zhang H.L. (1981) *Ultrasonics* **19,** 155. 160
Yuozonene L.V. (1980) *Sov. J. Nondestr. Test.* **16,** 574. 59

Name index

Index of symbols

Each entry contains a reference to the page on which the symbol is defined. Only usages which occur throughout the monograph are listed here: other usages are confined to individual sections and defined as they are required. The list is separated into Roman and Greek letters. Within each group vectors and tensors (bold symbols) are given first, followed by scalars (including elements of vectors and tensors).

S	surface, usually range of integration. 8
$t_i \equiv \mathbf{t}$	surface traction. 7
$u_i \equiv \mathbf{u}$	displacement under deformation. 6
U	internal energy density. 8
v_g	wave group velocity. 17
v_p	wave phase velocity. 17
V	volume, usually range of integration. 8
Y_{lm}	spherical harmonic. 74
Y_n	cylindrical Bessel function. 75
y_n	spherical Bessel function. 75
Y	Young's modulus. 10
Z	impedance. 19
$\boldsymbol{\omega} \equiv \omega_{ij}$	rotation tensor. 7
$\boldsymbol{\rho} \equiv \rho_i$	radius of curvature of wavefront. 115
$\boldsymbol{\Sigma} \equiv \Sigma_{ijk}$	stress Green function. 61
$\boldsymbol{\tau} \equiv \tau_{ij}$	stress tensor. 7
κ	ratio of shear to compression wave speed. 42
Λ	eigenfunction. 67
λ	Lamé elastic constant for an isotropic solid. 10
μ	Lamé elastic constant for an isotropic solid. 10
ν	Poisson's ratio. 10
$\omega_{ij} \equiv \boldsymbol{\omega}$	rotation tensor. 7
ω	angular frequency of wave. 16
Ω	dimensionless frequency of Lamb wave. 42
Ω	solid angle. 20
ϕ	compression wave potential. 14
ψ	shear wave potential. 14
$\rho_i \equiv \boldsymbol{\rho}$	radius of curvature of wavefront. 115
$\Sigma_{ijk} \equiv \boldsymbol{\Sigma}$	stress Green function. 61
σ	area of bundle of rays. 115
$\tau_{ij} \equiv \boldsymbol{\tau}$	stress tensor. 7
θ_c	critical angle 30
υ	dimensionless phase velocity of Lamb wave. 42
ξ	curvilinear coordinate. 66

General index